T0335643

PERIODIC
TABLE

Memorize the elements of the periodic table with its symbols and spelling

Yash'al Ahmed Abdul Sattar
Aminath Sharahath

Copyright © 2022 *Aminath Sharahath*. All rights reserved.

Cover Design by Fathmath Rizfa

All rights reserved. No part of this book may be used or reproduced by any means, graphic, electronic, or mechanical, including photocopying, recording, taping or by any information storage retrieval system without the written permission of the publisher except in the case of brief quotations embodied in critical articles and reviews.

To order additional copies of this book, contact
Toll Free +65 3165 7531 (Singapore)
Toll Free +60 3 3099 4412 (Malaysia)
www.partridgepublishing.com/singapore
orders.singapore@partridgepublishing.com

Because of the dynamic nature of the Internet, any web addresses or links contained in this book may have changed since publication and may no longer be valid. The views expressed in this work are solely those of the author and do not necessarily reflect the views of the publisher, and the publisher hereby disclaims any responsibility for them.

ISBN
978-1-5437-6975-3 (sc)
978-1-5437-6976-0 (hc)
978-1-5437-6977-7 (e)

Library of Congress Control Number: 2022907732

Print information available on the last page.

04/13/2022

PARTRIDGE

HOW TO USE THIS BOOK

Having the periodic table memorized along with chemical symbols and their spelling is going to be very helpful in your science class.

This book will help you memorize the periodic table, the symbols, and the spelling in a fun way. You won't even know you are learning. By the time you are done with the set of puzzles, you'll be amazed at how much you know within such a short period of time, while you are having fun!

The puzzles are very simple, and easy. It is divided into 12 groups according to the order in the periodic table. Each group will have 10 elements except the last group, which has 8 elements. There are 5 crossword puzzles in each group to help you with your learning process.

All you need to do is use the chemical symbols of the elements to fill in the crossword puzzle with the element name. The first 3 puzzles in each set will give you the word bank. They are color coded to help you. By the end of the first 3 puzzles, you will be familiar with the spelling so the word bank of the last 2 puzzles in the set will be replaced with a very simple fill in the blanks. They are color coded, so it is still very easy, but challenging enough to test your memory.

EXAMPLE 1 (filling the clues under 'DOWN')

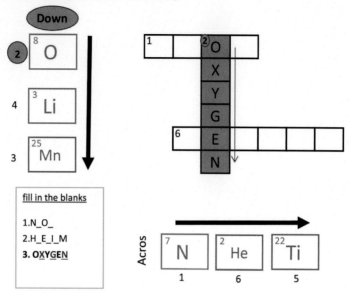

EXAMPLE 2 (filling the clues under 'ACROSS')

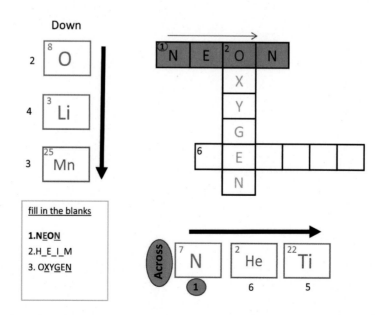

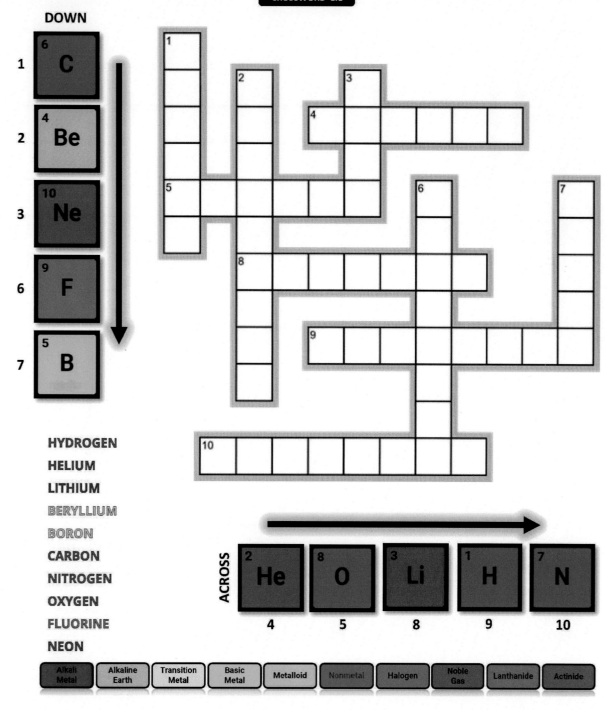

CROSSWORD 1.1

DOWN

1. C (6)
2. Be (4)
3. Ne (10)
6. F (9)
7. B (5)

HYDROGEN
HELIUM
LITHIUM
BERYLLIUM
BORON
CARBON
NITROGEN
OXYGEN
FLUORINE
NEON

ACROSS

He (2) — 4
O (8) — 5
Li (3) — 8
H (1) — 9
N (7) — 10

| Alkali Metal | Alkaline Earth | Transition Metal | Basic Metal | Metalloid | Nonmetal | Halogen | Noble Gas | Lanthanide | Actinide |

1

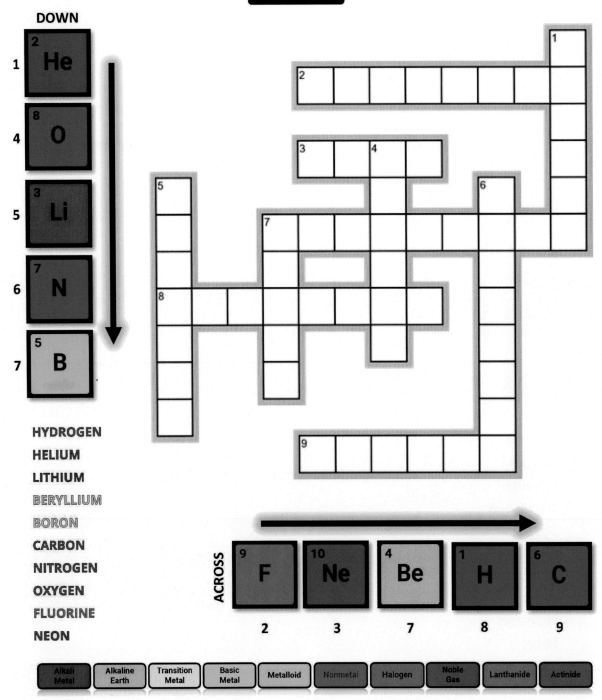

DOWN

1 He (2)

4 O (8)

5 Li (3)

6 N (7)

7 B (5)

HYDROGEN
HELIUM
LITHIUM
BERYLLIUM
BORON
CARBON
NITROGEN
OXYGEN
FLUORINE
NEON

ACROSS

2 F (9)

3 Ne (10)

7 Be (4)

8 H (1)

9 C (6)

Alkali Metal Alkaline Earth Transition Metal Basic Metal Metalloid Nonmetal Halogen Noble Gas Lanthanide Actinide

CROSSWORD 1.3

DOWN

2. **Li** (3)
3. **C** (6)
4. **H** (1)
5. **B** (5)
6. **F** (9)

HYDROGEN
HELIUM
LITHIUM
BERYLLIUM
BORON
CARBON
NITROGEN
OXYGEN
FLUORINE
NEON

ACROSS

1. **He** (2)
7. **Be** (4)
8. **Ne** (10)
9. **N** (7)
10. **O** (8)

| Alkali Metal | Alkaline Earth | Transition Metal | Basic Metal | Metalloid | Nonmetal | Halogen | Noble Gas | Lanthanide | Actinide |

CROSSWORD 1.4

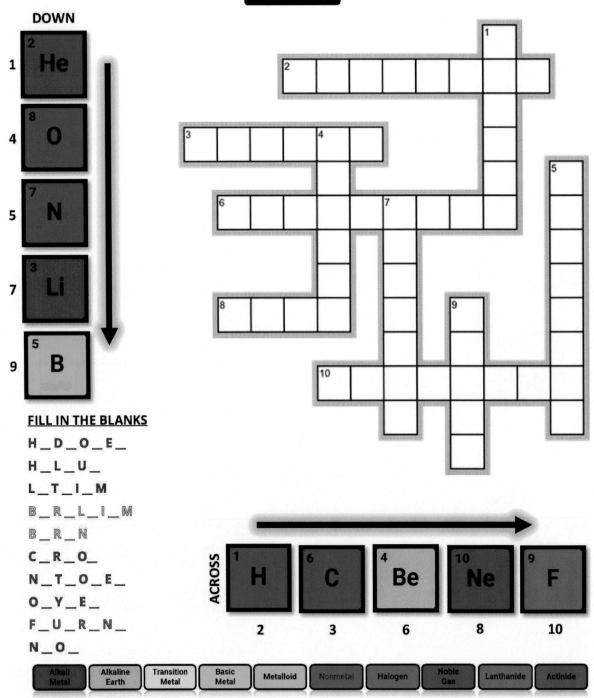

DOWN

1. **He** (2)
4. **O** (8)
5. **N** (7)
7. **Li** (3)
9. **B** (5)

FILL IN THE BLANKS

H _ D _ O _ E _
H _ L _ U _
L _ T _ I _ M
B _ R _ L _ I _ M
B _ R _ N
C _ R _ O _
N _ T _ O _ E _
O _ Y _ E _
F _ U _ R _ N _
N _ O _

ACROSS

2. **H** (1)
3. **C** (6)
6. **Be** (4)
8. **Ne** (10)
10. **F** (9)

Alkali Metal | Alkaline Earth | Transition Metal | Basic Metal | Metalloid | Nonmetal | Halogen | Noble Gas | Lanthanide | Actinide

4

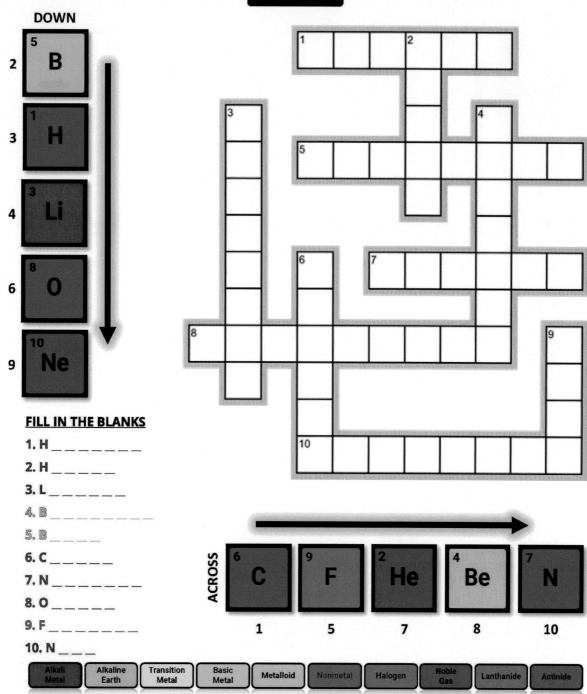

DOWN

2. B (5)
3. H (1)
4. Li (3)
6. O (8)
9. Ne (10)

FILL IN THE BLANKS

1. H _ _ _ _ _ _ _ _
2. H _ _ _ _ _
3. L _ _ _ _ _ _ _
4. B _ _ _ _ _ _ _ _
5. B _ _ _ _
6. C _ _ _ _ _ _
7. N _ _ _ _ _ _ _ _
8. O _ _ _ _ _
9. F _ _ _ _ _ _ _ _
10. N _ _ _

ACROSS

C (6) — 1
F (9) — 5
He (2) — 7
Be (4) — 8
N (7) — 10

Alkali Metal | Alkaline Earth | Transition Metal | Basic Metal | Metalloid | Nonmetal | Halogen | Noble Gas | Lanthanide | Actinide

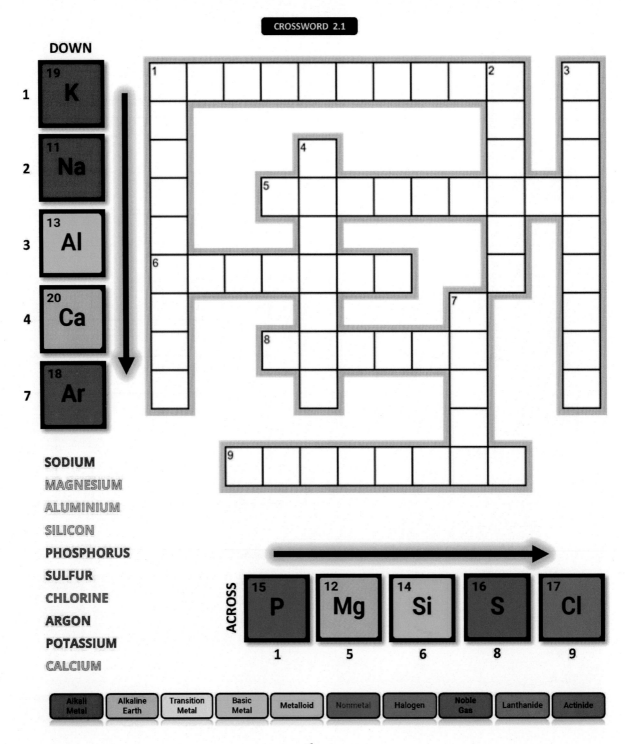

CROSSWORD 2.1

DOWN

1. K (19)
2. Na (11)
3. Al (13)
4. Ca (20)
7. Ar (18)

SODIUM
MAGNESIUM
ALUMINIUM
SILICON
PHOSPHORUS
SULFUR
CHLORINE
ARGON
POTASSIUM
CALCIUM

ACROSS

1. P (15)
5. Mg (12)
6. Si (14)
8. S (16)
9. Cl (17)

Alkali Metal | Alkaline Earth | Transition Metal | Basic Metal | Metalloid | Nonmetal | Halogen | Noble Gas | Lanthanide | Actinide

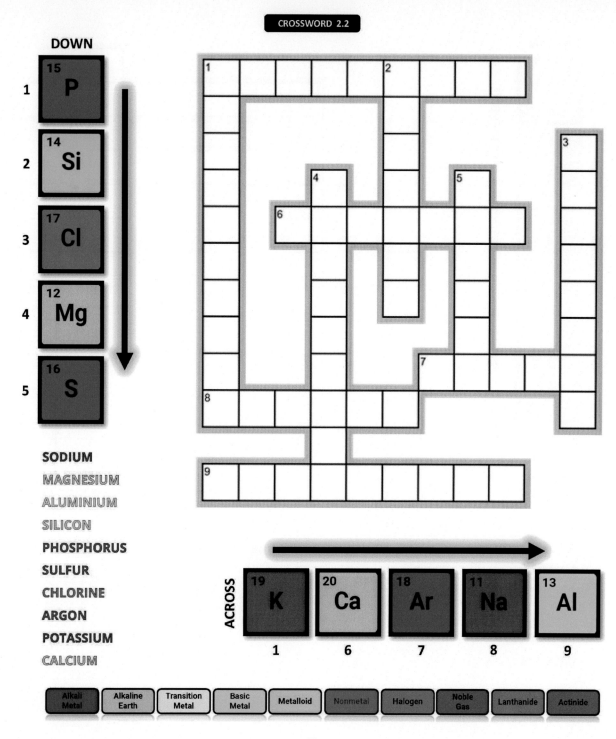

DOWN

1. 15 P

2. 14 Si

3. 17 Cl

4. 12 Mg

5. 16 S

SODIUM
MAGNESIUM
ALUMINIUM
SILICON
PHOSPHORUS
SULFUR
CHLORINE
ARGON
POTASSIUM
CALCIUM

ACROSS

1. 19 K
6. 20 Ca
7. 18 Ar
8. 11 Na
9. 13 Al

| Alkali Metal | Alkaline Earth | Transition Metal | Basic Metal | Metalloid | Nonmetal | Halogen | Noble Gas | Lanthanide | Actinide |

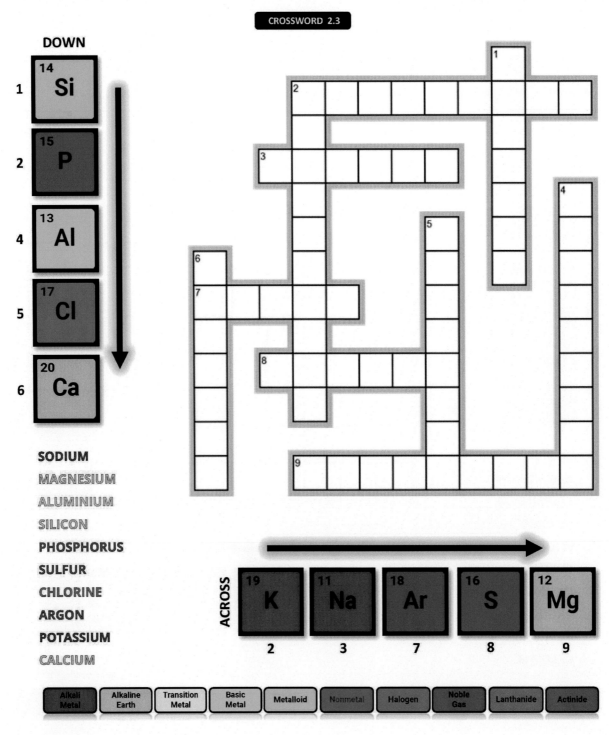

DOWN

1 | 14 Si
2 | 15 P
4 | 13 Al
5 | 17 Cl
6 | 20 Ca

SODIUM
MAGNESIUM
ALUMINIUM
SILICON
PHOSPHORUS
SULFUR
CHLORINE
ARGON
POTASSIUM
CALCIUM

ACROSS

19 K	11 Na	18 Ar	16 S	12 Mg
2	3	7	8	9

Alkali Metal | Alkaline Earth | Transition Metal | Basic Metal | Metalloid | Nonmetal | Halogen | Noble Gas | Lanthanide | Actinide

DOWN

1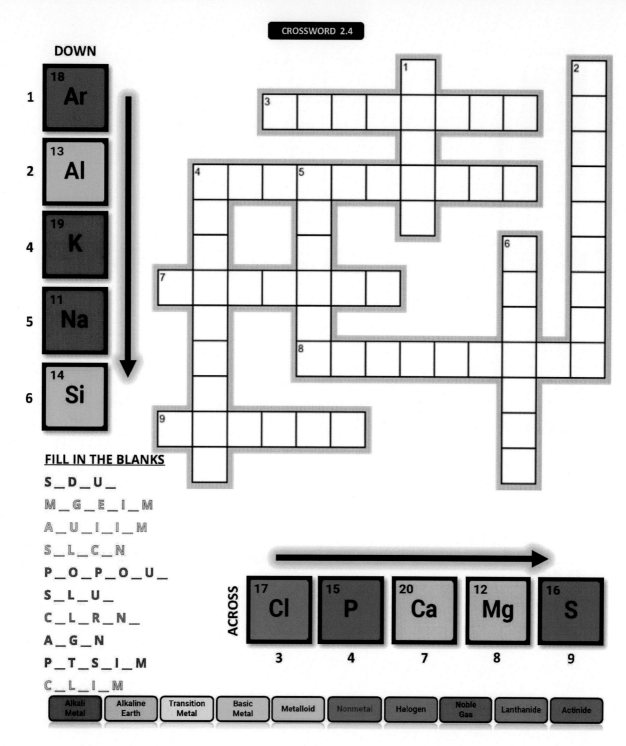

18	Ar

1

13	Al

2

19	K

4

11	Na

5

14	Si

6

FILL IN THE BLANKS

S _ D _ U _

M _ G _ E _ I _ M

A _ U _ I _ I _ M

S _ L _ C _ N

P _ O _ P _ O _ U _

S _ L _ U _

C _ L _ R _ N _

A _ G _ N

P _ T _ S _ I _ M

C _ L _ I _ M

ACROSS

17	Cl	15	P	20	Ca	12	Mg	16	S

3   4   7   8   9

| Alkali Metal | Alkaline Earth | Transition Metal | Basic Metal | Metalloid | Nonmetal | Halogen | Noble Gas | Lanthanide | Actinide |

9

CROSSWORD 2.5

DOWN

2 | 18 Ar
3 | 17 Cl
4 | 13 Al
6 | 11 Na
7 | 20 Ca

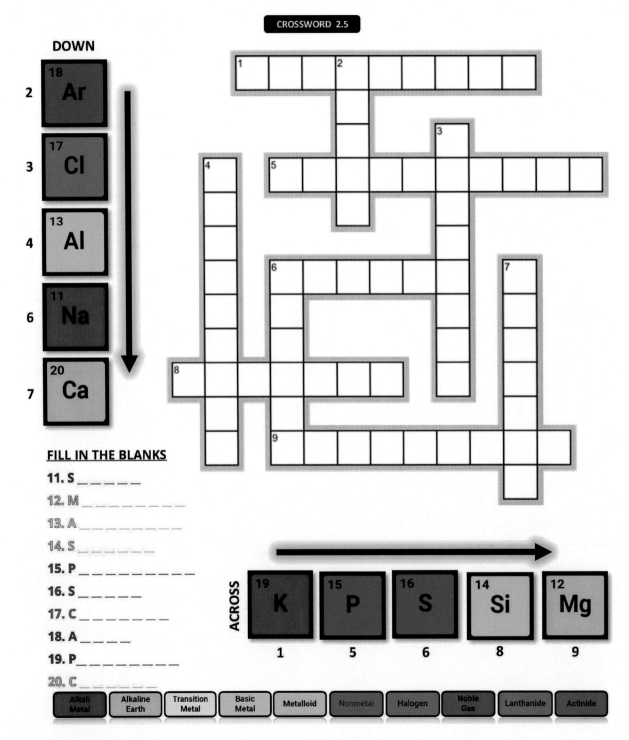

FILL IN THE BLANKS

11. S _ _ _ _ _ _

12. M _ _ _ _ _ _ _ _ _

13. A _ _ _ _ _ _ _

14. S _ _ _ _ _ _

15. P _ _ _ _ _ _ _ _ _ _

16. S _ _ _ _ _

17. C _ _ _ _ _ _ _ _

18. A _ _ _ _ _

19. P _ _ _ _ _ _ _ _ _

20. C _ _ _ _ _ _ _

ACROSS

19 K	15 P	16 S	14 Si	12 Mg
1	5	6	8	9

| Alkali Metal | Alkaline Earth | Transition Metal | Basic Metal | Metalloid | Nonmetal | Halogen | Noble Gas | Lanthanide | Actinide |

10

CROSSWORD 3.1

DOWN

#		
1	23	V
3	27	Co
4	25	Mn
5	24	Cr
8	30	Zn

SCANDIUM
TITANIUM
VANADIUM
CHROMIUM
MANGANESE
IRON
COBALT
NICKEL
COPPER
ZINC

ACROSS

21 Sc	26 Fe	22 Ti	28 Ni	29 Cu
2	6	7	9	10

Alkali Metal | Alkaline Earth | Transition Metal | Basic Metal | Metalloid | Nonmetal | Halogen | Noble Gas | Lanthanide | Actinide

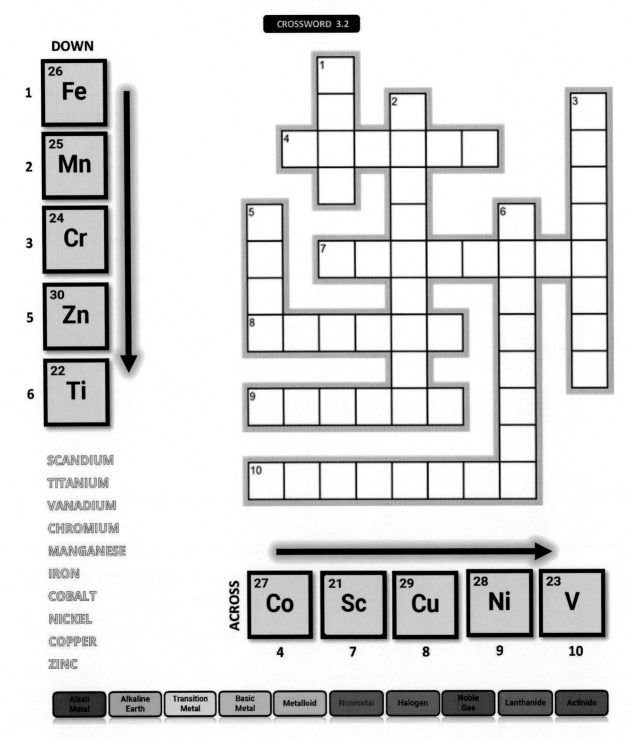

CROSSWORD 3.2

DOWN

#		
1	26	Fe
2	25	Mn
3	24	Cr
5	30	Zn
6	22	Ti

SCANDIUM
TITANIUM
VANADIUM
CHROMIUM
MANGANESE
IRON
COBALT
NICKEL
COPPER
ZINC

ACROSS

27 Co	21 Sc	29 Cu	28 Ni	23 V
4	7	8	9	10

| Alkali Metal | Alkaline Earth | Transition Metal | Basic Metal | Metalloid | Nonmetal | Halogen | Noble Gas | Lanthanide | Actinide |

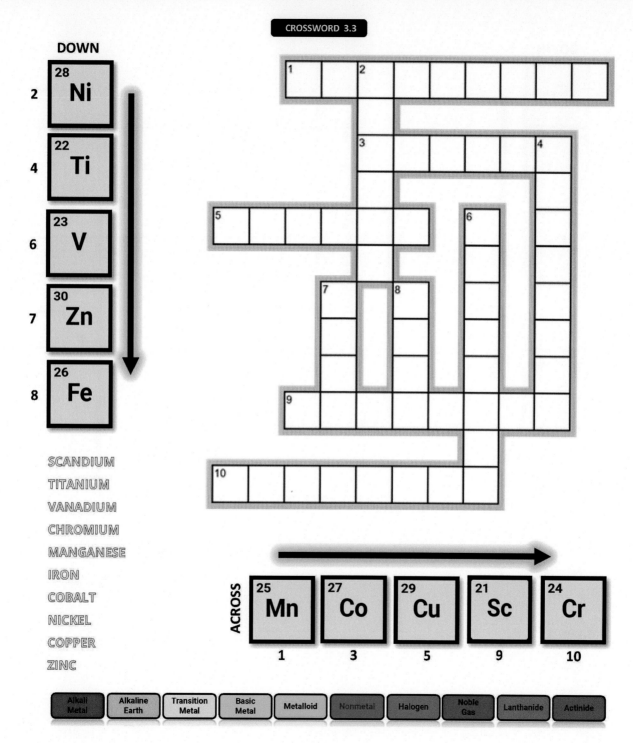

CROSSWORD 3.3

DOWN

#	Atomic Number	Symbol
2	28	Ni
4	22	Ti
6	23	V
7	30	Zn
8	26	Fe

SCANDIUM
TITANIUM
VANADIUM
CHROMIUM
MANGANESE
IRON
COBALT
NICKEL
COPPER
ZINC

ACROSS

#	Atomic Number	Symbol
1	25	Mn
3	27	Co
5	29	Cu
9	21	Sc
10	24	Cr

Alkali Metal · Alkaline Earth · Transition Metal · Basic Metal · Metalloid · Nonmetal · Halogen · Noble Gas · Lanthanide · Actinide

13

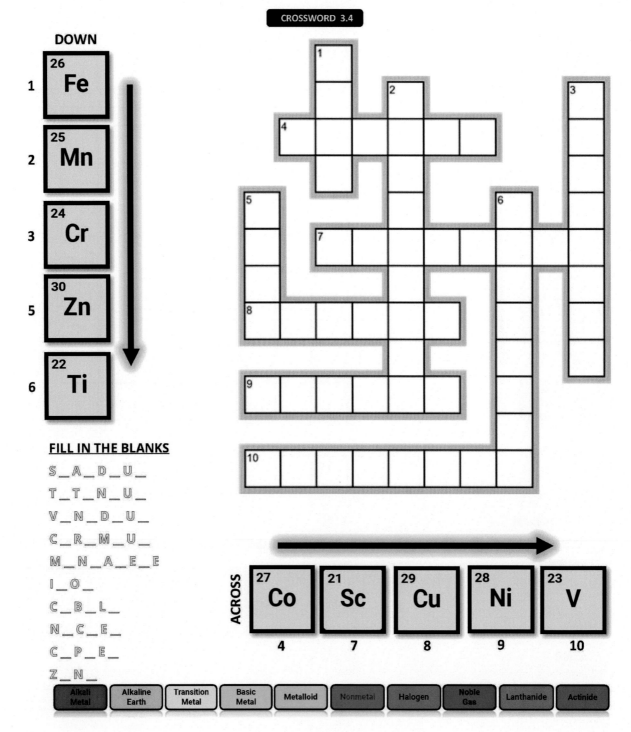

CROSSWORD 3.4

DOWN

1. ²⁶ Fe
2. ²⁵ Mn
3. ²⁴ Cr
5. ³⁰ Zn
6. ²² Ti

FILL IN THE BLANKS

S _ A _ D _ U _
T _ T _ N _ U _
V _ N _ D _ U _
C _ R _ M _ U _
M _ N _ A _ E _ E
I _ O _
C _ B _ L _
N _ C _ E _
C _ P _ E _
Z _ N _

ACROSS

4. ²⁷ Co
7. ²¹ Sc
8. ²⁹ Cu
9. ²⁸ Ni
10. ²³ V

| Alkali Metal | Alkaline Earth | Transition Metal | Basic Metal | Metalloid | Nonmetal | Halogen | Noble Gas | Lanthanide | Actinide |

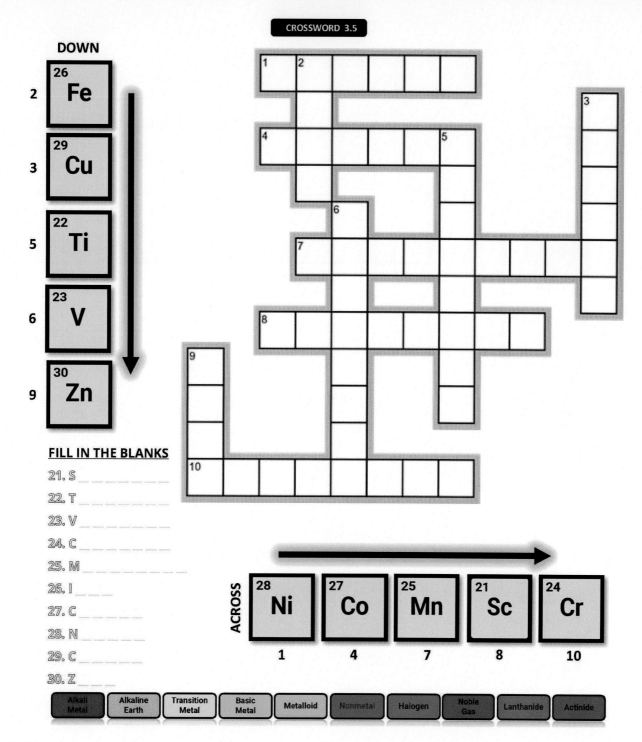

CROSSWORD 3.5

DOWN

2. 26 Fe
3. 29 Cu
5. 22 Ti
6. 23 V
9. 30 Zn

FILL IN THE BLANKS

21. S _____
22. T _____
23. V _____
24. C _____
25. M _____
26. I _____
27. C _____
28. N _____
29. C _____
30. Z _____

ACROSS

1. 28 Ni
4. 27 Co
7. 25 Mn
8. 21 Sc
10. 24 Cr

Alkali Metal | Alkaline Earth | Transition Metal | Basic Metal | Metalloid | Nonmetal | Halogen | Noble Gas | Lanthanide | Actinide

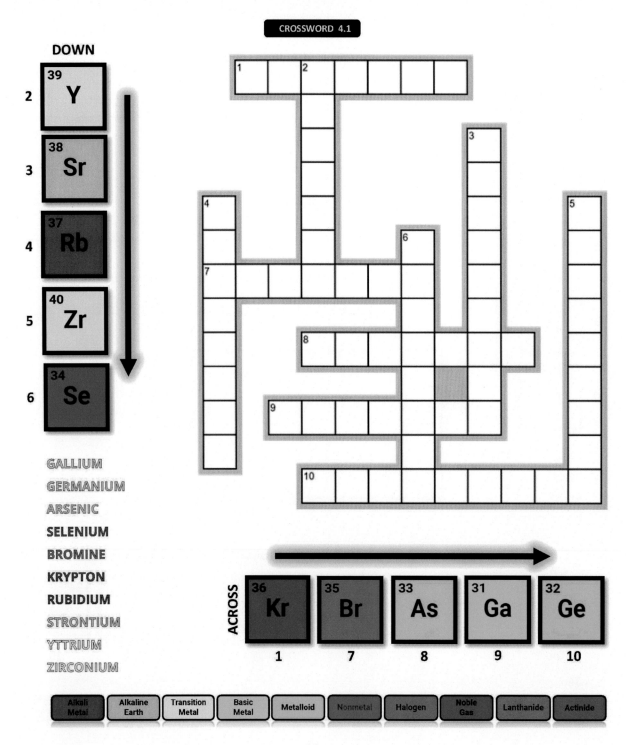

CROSSWORD 4.1

DOWN

2 — 39 Y
3 — 38 Sr
4 — 37 Rb
5 — 40 Zr
6 — 34 Se

GALLIUM
GERMANIUM
ARSENIC
SELENIUM
BROMINE
KRYPTON
RUBIDIUM
STRONTIUM
YTTRIUM
ZIRCONIUM

ACROSS

36 Kr — 1
35 Br — 7
33 As — 8
31 Ga — 9
32 Ge — 10

| Alkali Metal | Alkaline Earth | Transition Metal | Basic Metal | Metalloid | Nonmetal | Halogen | Noble Gas | Lanthanide | Actinide |

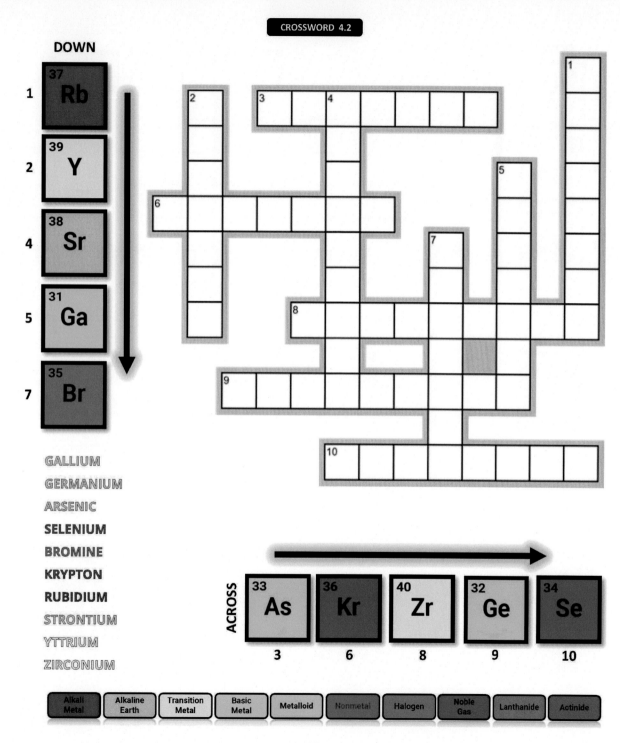

CROSSWORD 4.2

DOWN

1 [37 Rb]
2 [39 Y]
4 [38 Sr]
5 [31 Ga]
7 [35 Br]

GALLIUM
GERMANIUM
ARSENIC
SELENIUM
BROMINE
KRYPTON
RUBIDIUM
STRONTIUM
YTTRIUM
ZIRCONIUM

ACROSS

[33 As] 3
[36 Kr] 6
[40 Zr] 8
[32 Ge] 9
[34 Se] 10

| Alkali Metal | Alkaline Earth | Transition Metal | Basic Metal | Metalloid | Nonmetal | Halogen | Noble Gas | Lanthanide | Actinide |

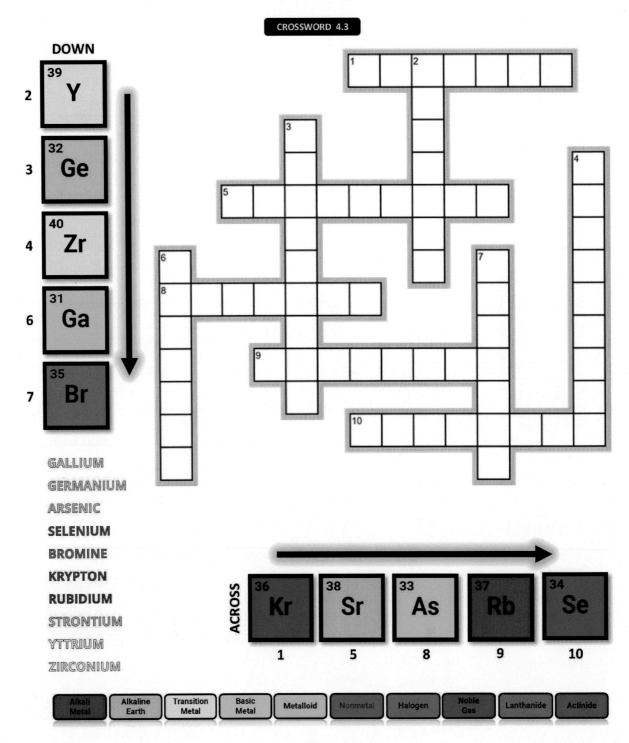

CROSSWORD 4.3

DOWN

2 | **39** Y
3 | **32** Ge
4 | **40** Zr
6 | **31** Ga
7 | **35** Br

GALLIUM
GERMANIUM
ARSENIC
SELENIUM
BROMINE
KRYPTON
RUBIDIUM
STRONTIUM
YTTRIUM
ZIRCONIUM

ACROSS

| **36** Kr | **38** Sr | **33** As | **37** Rb | **34** Se |
| 1 | 5 | 8 | 9 | 10 |

| Alkali Metal | Alkaline Earth | Transition Metal | Basic Metal | Metalloid | Nonmetal | Halogen | Noble Gas | Lanthanide | Actinide |

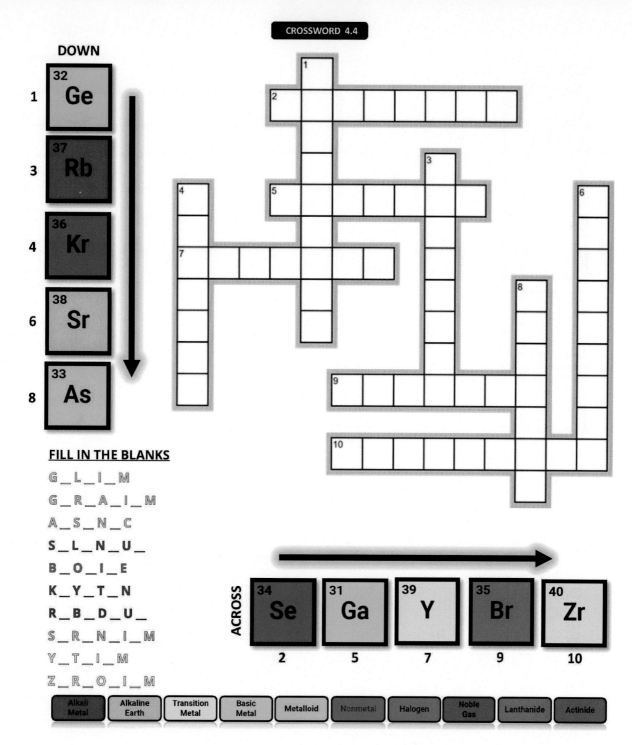

CROSSWORD 4.4

DOWN

1. **32 Ge**
3. **37 Rb**
4. **36 Kr**
6. **38 Sr**
8. **33 As**

FILL IN THE BLANKS

G _ L _ I _ M

G _ R _ A _ I _ M

A _ S _ N _ C

S _ L _ N _ U _

B _ O _ I _ E

K _ Y _ T _ N

R _ B _ D _ U _

S _ R _ N _ I _ M

Y _ T _ I _ M

Z _ R _ O _ I _ M

ACROSS

2	5	7	9	10
34 Se	**31 Ga**	**39 Y**	**35 Br**	**40 Zr**

Alkali Metal | Alkaline Earth | Transition Metal | Basic Metal | Metalloid | Nonmetal | Halogen | Noble Gas | Lanthanide | Actinide

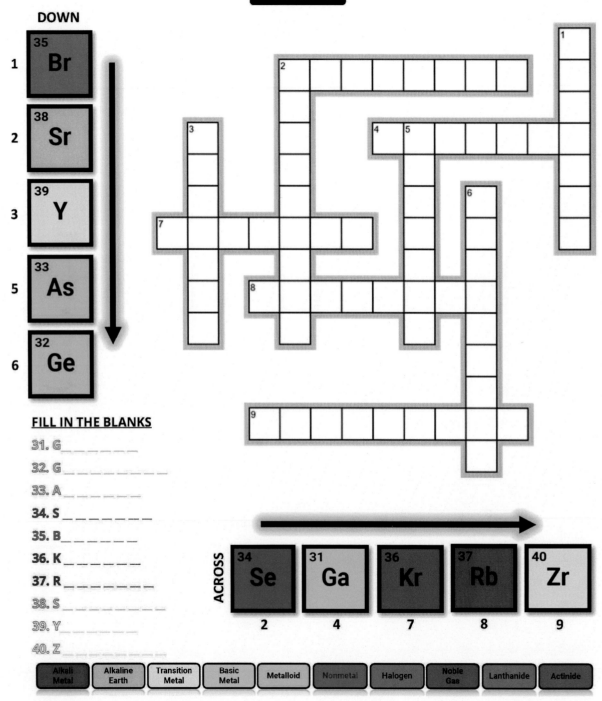

CROSSWORD 4.5

DOWN

1. ^{35} Br
2. ^{38} Sr
3. ^{39} Y
5. ^{33} As
6. ^{32} Ge

FILL IN THE BLANKS

31. G _ _ _ _ _ _ _
32. G _ _ _ _ _ _ _ _
33. A _ _ _ _ _ _
34. S _ _ _ _ _ _ _ _
35. B _ _ _ _ _ _
36. K _ _ _ _ _ _ _
37. R _ _ _ _ _ _ _
38. S _ _ _ _ _ _ _
39. Y _ _ _ _ _ _
40. Z _ _ _ _ _ _ _ _

ACROSS

2. ^{34} Se
4. ^{31} Ga
7. ^{36} Kr
8. ^{37} Rb
9. ^{40} Zr

Alkali Metal | Alkaline Earth | Transition Metal | Basic Metal | Metalloid | Nonmetal | Halogen | Noble Gas | Lanthanide | Actinide

20

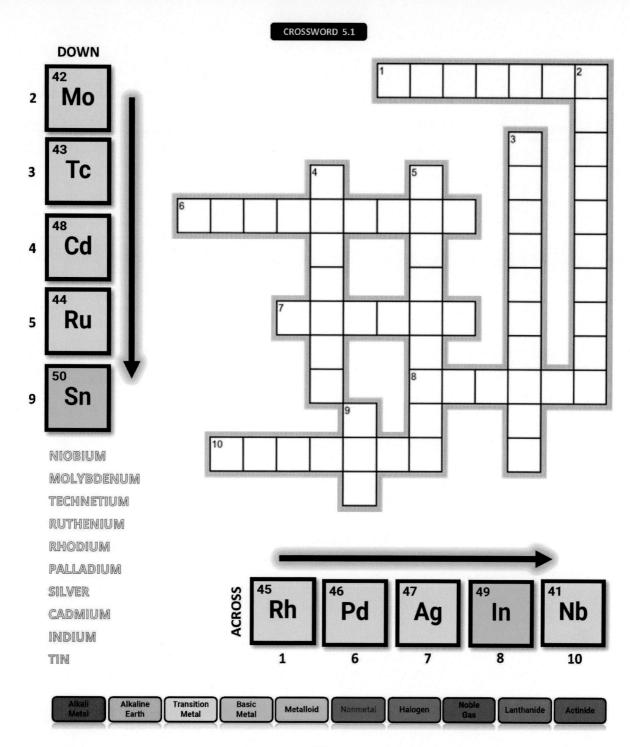

CROSSWORD 5.1

DOWN

2 — 42 Mo
3 — 43 Tc
4 — 48 Cd
5 — 44 Ru
9 — 50 Sn

NIOBIUM
MOLYBDENUM
TECHNETIUM
RUTHENIUM
RHODIUM
PALLADIUM
SILVER
CADMIUM
INDIUM
TIN

ACROSS

45 Rh — 1
46 Pd — 6
47 Ag — 7
49 In — 8
41 Nb — 10

| Alkali Metal | Alkaline Earth | Transition Metal | Basic Metal | Metalloid | Nonmetal | Halogen | Noble Gas | Lanthanide | Actinide |

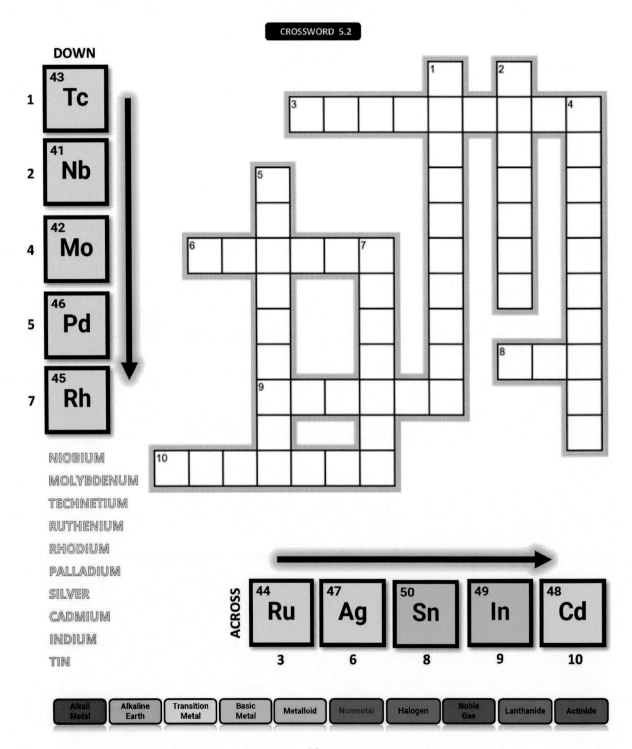

CROSSWORD 5.2

DOWN

1 **Tc** 43
2 **Nb** 41
4 **Mo** 42
5 **Pd** 46
7 **Rh** 45

NIOBIUM
MOLYBDENUM
TECHNETIUM
RUTHENIUM
RHODIUM
PALLADIUM
SILVER
CADMIUM
INDIUM
TIN

ACROSS

Ru 44 — 3
Ag 47 — 6
Sn 50 — 8
In 49 — 9
Cd 48 — 10

| Alkali Metal | Alkaline Earth | Transition Metal | Basic Metal | Metalloid | Nonmetal | Halogen | Noble Gas | Lanthanide | Actinide |

22

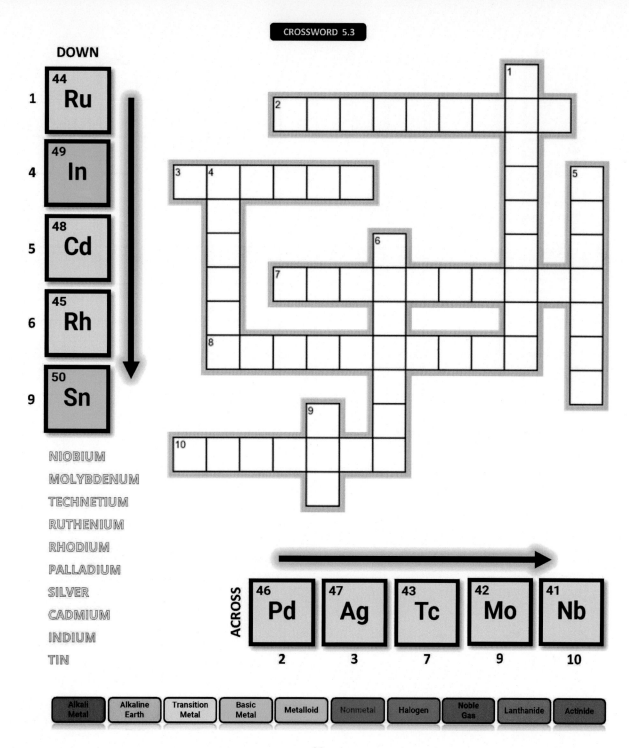

CROSSWORD 5.3

DOWN

#	Z	Symbol
1	44	Ru
4	49	In
5	48	Cd
6	45	Rh
9	50	Sn

NIOBIUM
MOLYBDENUM
TECHNETIUM
RUTHENIUM
RHODIUM
PALLADIUM
SILVER
CADMIUM
INDIUM
TIN

ACROSS

46	47	43	42	41
Pd	Ag	Tc	Mo	Nb
2	3	7	9	10

Alkali Metal | Alkaline Earth | Transition Metal | Basic Metal | Metalloid | Nonmetal | Halogen | Noble Gas | Lanthanide | Actinide

23

DOWN

1. 43 Tc
2. 41 Nb
4. 42 Mo
5. 44 Ru
7. 50 Sn

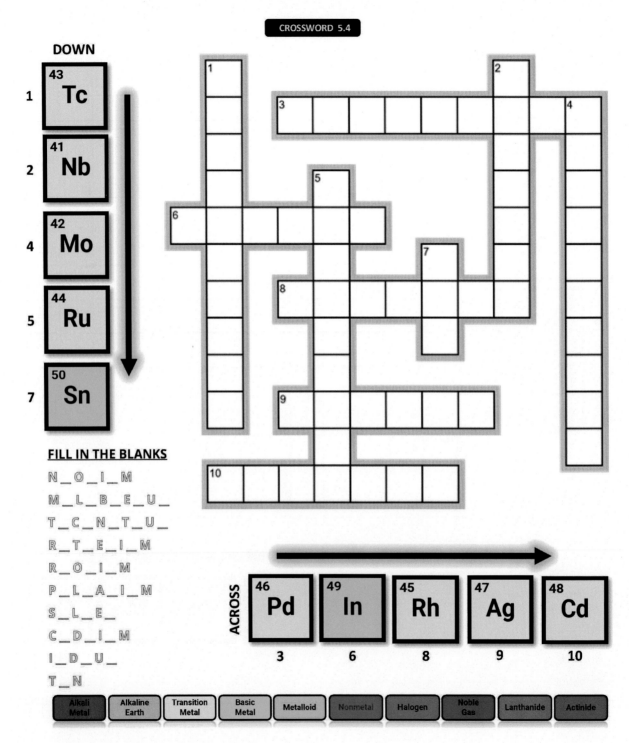

FILL IN THE BLANKS

N _ _ O _ I _ M

M _ L _ B _ E _ U _

T _ C _ N _ T _ U _

R _ T _ E _ I _ M

R _ O _ I _ M

P _ L _ A _ I _ M

S _ L _ E _

C _ D _ I _ M

I _ D _ U _

T _ N

ACROSS

46 Pd — 3
49 In — 6
45 Rh — 8
47 Ag — 9
48 Cd — 10

Alkali Metal | Alkaline Earth | Transition Metal | Basic Metal | Metalloid | Nonmetal | Halogen | Noble Gas | Lanthanide | Actinide

24

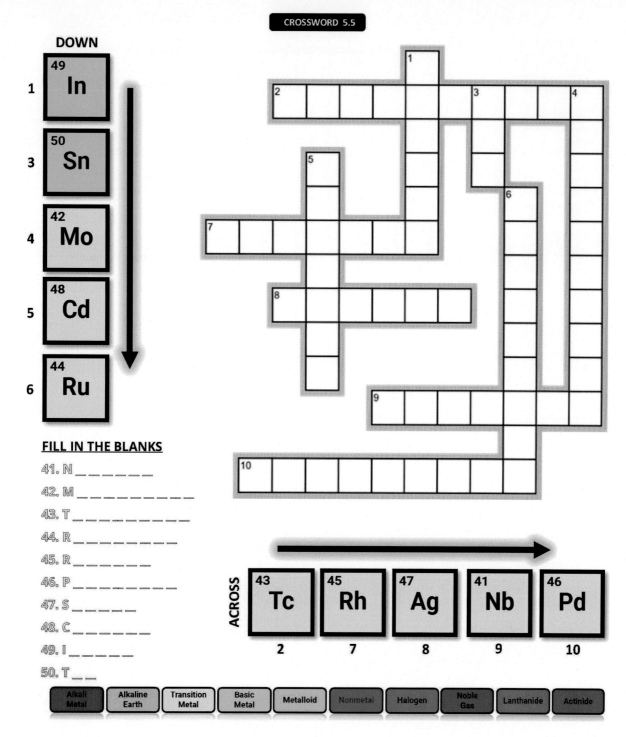

CROSSWORD 5.5

DOWN

1. 49 **In**
3. 50 **Sn**
4. 42 **Mo**
5. 48 **Cd**
6. 44 **Ru**

FILL IN THE BLANKS

41. N _ _ _ _ _ _ _
42. M _ _ _ _ _ _ _ _ _
43. T _ _ _ _ _ _ _ _ _
44. R _ _ _ _ _ _ _ _
45. R _ _ _ _ _ _ _
46. P _ _ _ _ _ _ _ _ _
47. S _ _ _ _ _ _
48. C _ _ _ _ _ _ _
49. I _ _ _ _ _ _
50. T _ _ _

ACROSS

43 Tc	45 Rh	47 Ag	41 Nb	46 Pd
2	7	8	9	10

Alkali Metal | Alkaline Earth | Transition Metal | Basic Metal | Metalloid | Nonmetal | Halogen | Noble Gas | Lanthanide | Actinide

25

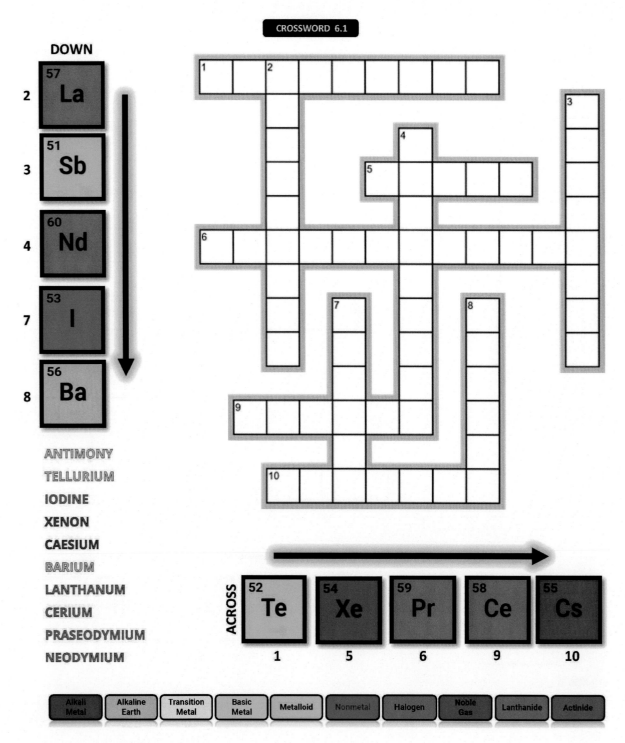

CROSSWORD 6.1

DOWN

2 — 57 La
3 — 51 Sb
4 — 60 Nd
7 — 53 I
8 — 56 Ba

ANTIMONY
TELLURIUM
IODINE
XENON
CAESIUM
BARIUM
LANTHANUM
CERIUM
PRASEODYMIUM
NEODYMIUM

ACROSS

1 — 52 Te
5 — 54 Xe
6 — 59 Pr
9 — 58 Ce
10 — 55 Cs

Alkali Metal | Alkaline Earth | Transition Metal | Basic Metal | Metalloid | Nonmetal | Halogen | Noble Gas | Lanthanide | Actinide

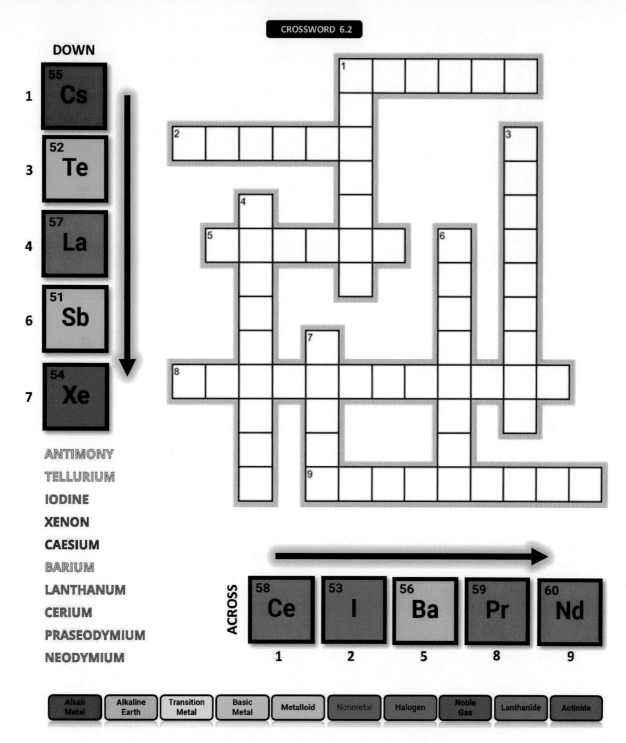

CROSSWORD 6.2

DOWN

1. 55 Cs
3. 52 Te
4. 57 La
6. 51 Sb
7. 54 Xe

ANTIMONY
TELLURIUM
IODINE
XENON
CAESIUM
BARIUM
LANTHANUM
CERIUM
PRASEODYMIUM
NEODYMIUM

ACROSS

1. 58 Ce
2. 53 I
5. 56 Ba
8. 59 Pr
9. 60 Nd

Alkali Metal | Alkaline Earth | Transition Metal | Basic Metal | Metalloid | Nonmetal | Halogen | Noble Gas | Lanthanide | Actinide

27

CROSSWORD 6.3

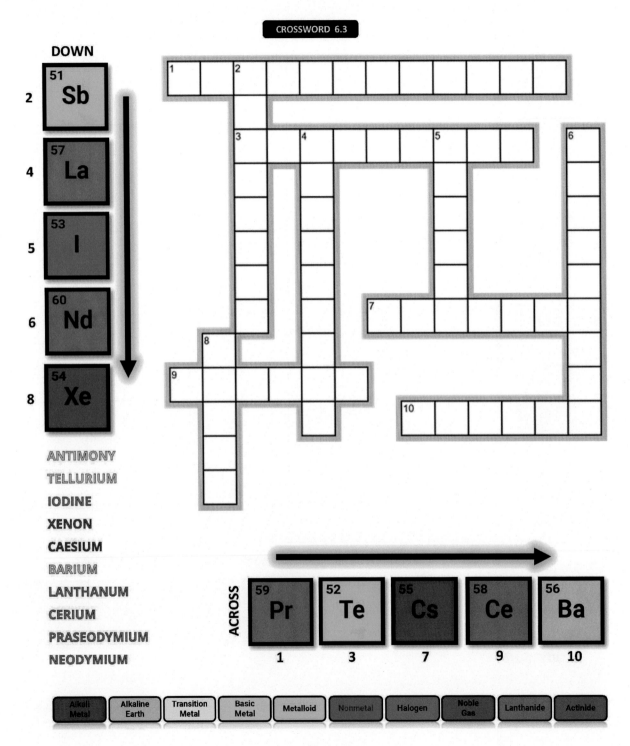

DOWN

2 — 51 Sb
4 — 57 La
5 — 53 I
6 — 60 Nd
8 — 54 Xe

ANTIMONY
TELLURIUM
IODINE
XENON
CAESIUM
BARIUM
LANTHANUM
CERIUM
PRASEODYMIUM
NEODYMIUM

ACROSS

59 Pr — 1
52 Te — 3
55 Cs — 7
58 Ce — 9
56 Ba — 10

Alkali Metal | Alkaline Earth | Transition Metal | Basic Metal | Metalloid | Nonmetal | Halogen | Noble Gas | Lanthanide | Actinide

DOWN

2 | 57 **La**
3 | 51 **Sb**
4 | 56 **Ba**
6 | 58 **Ce**
8 | 53 **I**

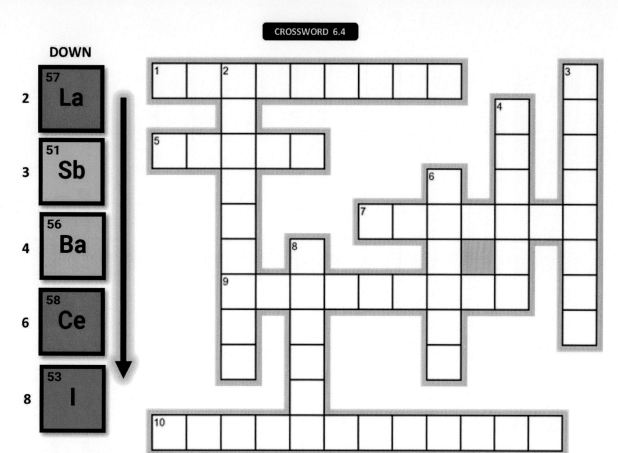

FILL IN THE BLANKS

A _ T _ M _ N _
T _ L _ U _ I _ M
I _ D _ N _
X _ N _ N
C _ E _ I _ M
B _ R _ U _
L _ N _ H _ N _ M
C _ R _ U _
P _ A _ E _ D _ M _ U _
N _ O _ Y _ I _ M

ACROSS

52 **Te**	54 **Xe**	55 **Cs**	60 **Nd**	59 **Pr**
1	5	7	9	10

| Alkali Metal | Alkaline Earth | Transition Metal | Basic Metal | Metalloid | Nonmetal | Halogen | Noble Gas | Lanthanide | Actinide |

29

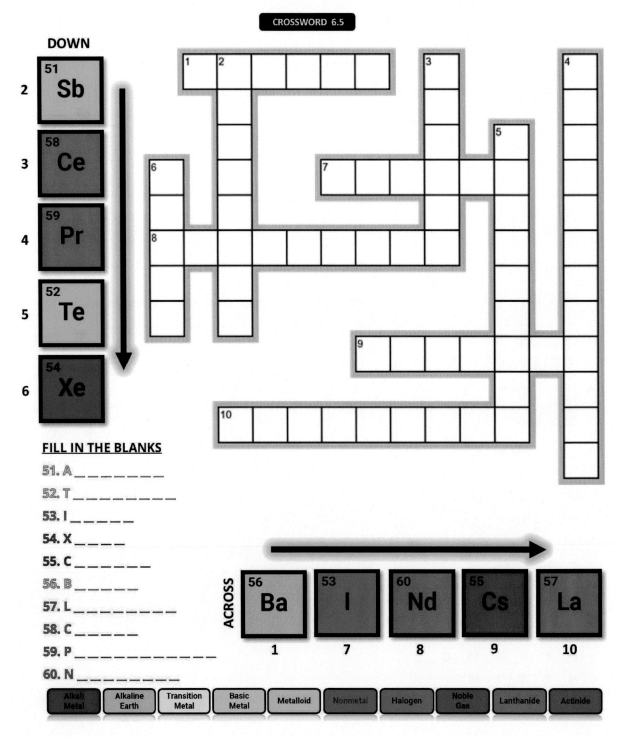

CROSSWORD 6.5

DOWN

2 | 51 Sb
3 | 58 Ce
4 | 59 Pr
5 | 52 Te
6 | 54 Xe

FILL IN THE BLANKS

51. A _ _ _ _ _ _ _ _

52. T _ _ _ _ _ _ _ _ _

53. I _ _ _ _ _ _

54. X _ _ _ _ _

55. C _ _ _ _ _ _ _

56. B _ _ _ _ _ _

57. L _ _ _ _ _ _ _ _ _

58. C _ _ _ _ _ _

59. P _ _ _ _ _ _ _ _ _ _ _ _ _

60. N _ _ _ _ _ _ _ _ _

ACROSS

56 Ba — 1
53 I — 7
60 Nd — 8
55 Cs — 9
57 La — 10

Alkali Metal | Alkaline Earth | Transition Metal | Basic Metal | Metalloid | Nonmetal | Halogen | Noble Gas | Lanthanide | Actinide

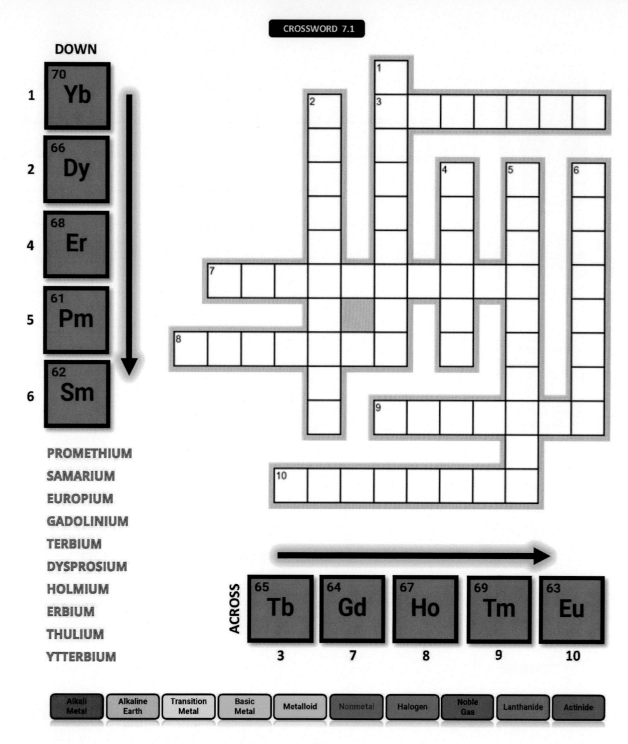

CROSSWORD 7.1

DOWN

1. ⁷⁰ Yb
2. ⁶⁶ Dy
4. ⁶⁸ Er
5. ⁶¹ Pm
6. ⁶² Sm

PROMETHIUM
SAMARIUM
EUROPIUM
GADOLINIUM
TERBIUM
DYSPROSIUM
HOLMIUM
ERBIUM
THULIUM
YTTERBIUM

ACROSS

⁶⁵ Tb — 3
⁶⁴ Gd — 7
⁶⁷ Ho — 8
⁶⁹ Tm — 9
⁶³ Eu — 10

Alkali Metal | Alkaline Earth | Transition Metal | Basic Metal | Metalloid | Nonmetal | Halogen | Noble Gas | Lanthanide | Actinide

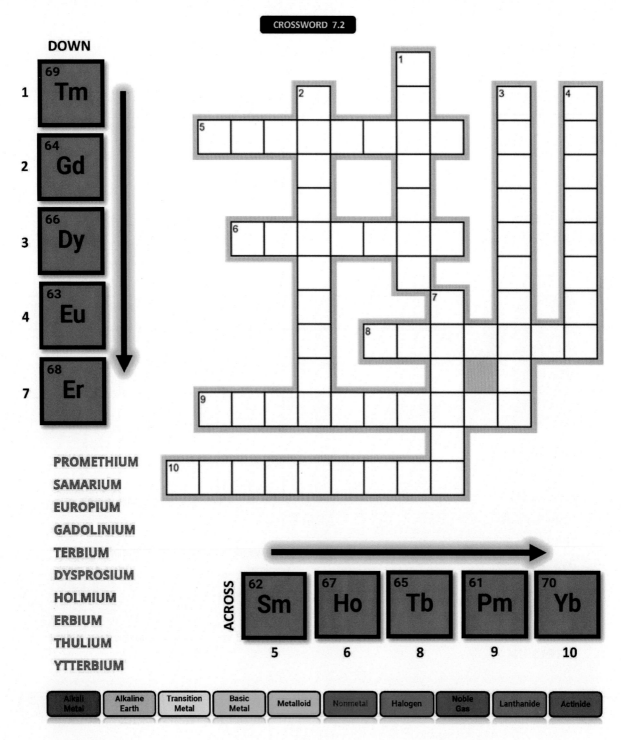

CROSSWORD 7.2

DOWN

1. ⁶⁹ Tm
2. ⁶⁴ Gd
3. ⁶⁶ Dy
4. ⁶³ Eu
7. ⁶⁸ Er

PROMETHIUM
SAMARIUM
EUROPIUM
GADOLINIUM
TERBIUM
DYSPROSIUM
HOLMIUM
ERBIUM
THULIUM
YTTERBIUM

ACROSS

5. ⁶² Sm
6. ⁶⁷ Ho
8. ⁶⁵ Tb
9. ⁶¹ Pm
10. ⁷⁰ Yb

Alkali Metal | Alkaline Earth | Transition Metal | Basic Metal | Metalloid | Nonmetal | Halogen | Noble Gas | Lanthanide | Actinide

32

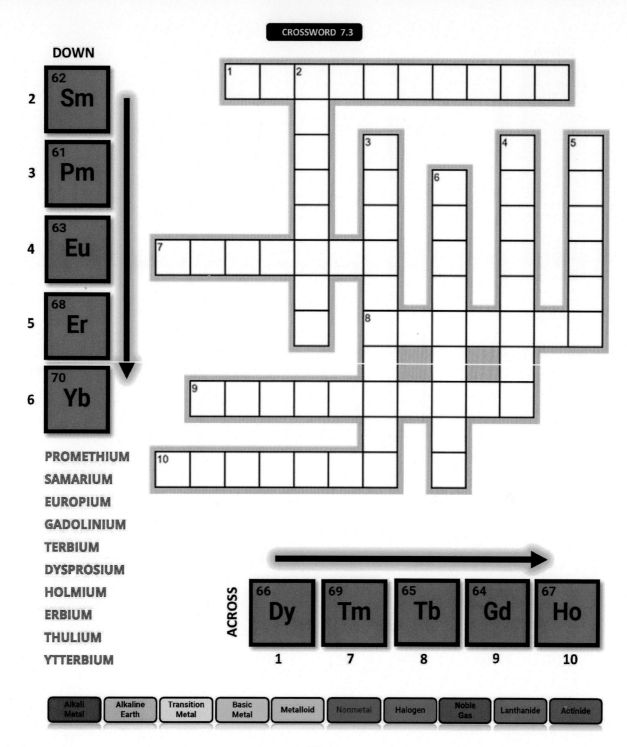

CROSSWORD 7.3

DOWN

2 Sm 62

3 Pm 61

4 Eu 63

5 Er 68

6 Yb 70

PROMETHIUM
SAMARIUM
EUROPIUM
GADOLINIUM
TERBIUM
DYSPROSIUM
HOLMIUM
ERBIUM
THULIUM
YTTERBIUM

ACROSS

1 Dy 66

7 Tm 69

8 Tb 65

9 Gd 64

10 Ho 67

Alkali Metal | Alkaline Earth | Transition Metal | Basic Metal | Metalloid | Nonmetal | Halogen | Noble Gas | Lanthanide | Actinide

33

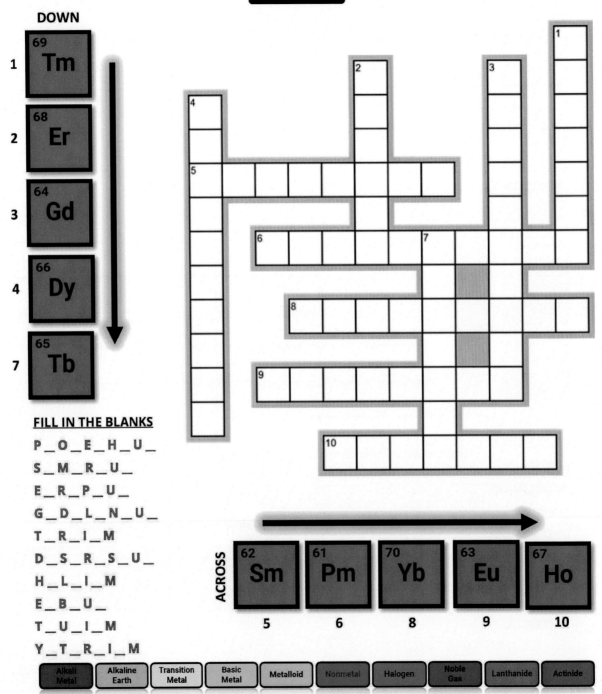

DOWN

1 69 Tm

2 68 Er

3 64 Gd

4 66 Dy

7 65 Tb

FILL IN THE BLANKS

P _ O _ E _ H _ U _

S _ M _ R _ U _

E _ R _ P _ U _

G _ D _ L _ N _ U _

T _ R _ I _ M

D _ S _ R _ S _ U _

H _ L _ I _ M

E _ B _ U _

T _ U _ I _ M

Y _ T _ R _ I _ M

ACROSS

62 Sm — 5
61 Pm — 6
70 Yb — 8
63 Eu — 9
67 Ho — 10

Alkali Metal | Alkaline Earth | Transition Metal | Basic Metal | Metalloid | Nonmetal | Halogen | Noble Gas | Lanthanide | Actinide

DOWN

1. 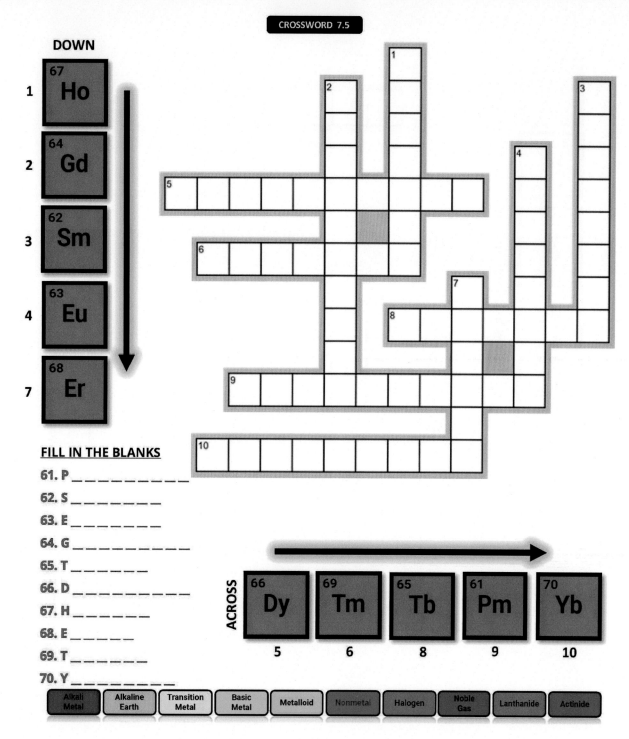 67 Ho

2. 64 Gd

3. 62 Sm

4. 63 Eu

7. 68 Er

FILL IN THE BLANKS

61. P _ _ _ _ _ _ _ _ _ _

62. S _ _ _ _ _ _ _ _

63. E _ _ _ _ _ _ _ _

64. G _ _ _ _ _ _ _ _ _ _

65. T _ _ _ _ _ _ _

66. D _ _ _ _ _ _ _ _ _ _

67. H _ _ _ _ _ _

68. E _ _ _ _ _ _

69. T _ _ _ _ _ _ _

70. Y _ _ _ _ _ _ _ _ _

ACROSS

66 Dy	69 Tm	65 Tb	61 Pm	70 Yb
5	6	8	9	10

Alkali Metal | Alkaline Earth | Transition Metal | Basic Metal | Metalloid | Nonmetal | Halogen | Noble Gas | Lanthanide | Actinide

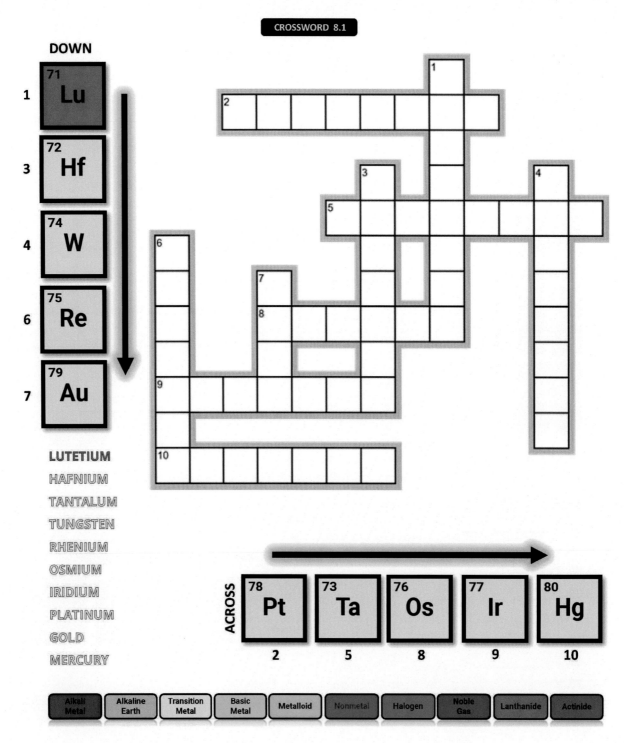

CROSSWORD 8.1

DOWN

1. Lu (71)
3. Hf (72)
4. W (74)
6. Re (75)
7. Au (79)

LUTETIUM
HAFNIUM
TANTALUM
TUNGSTEN
RHENIUM
OSMIUM
IRIDIUM
PLATINUM
GOLD
MERCURY

ACROSS

2. Pt (78)
5. Ta (73)
8. Os (76)
9. Ir (77)
10. Hg (80)

| Alkali Metal | Alkaline Earth | Transition Metal | Basic Metal | Metalloid | Nonmetal | Halogen | Noble Gas | Lanthanide | Actinide |

CROSSWORD 8.2

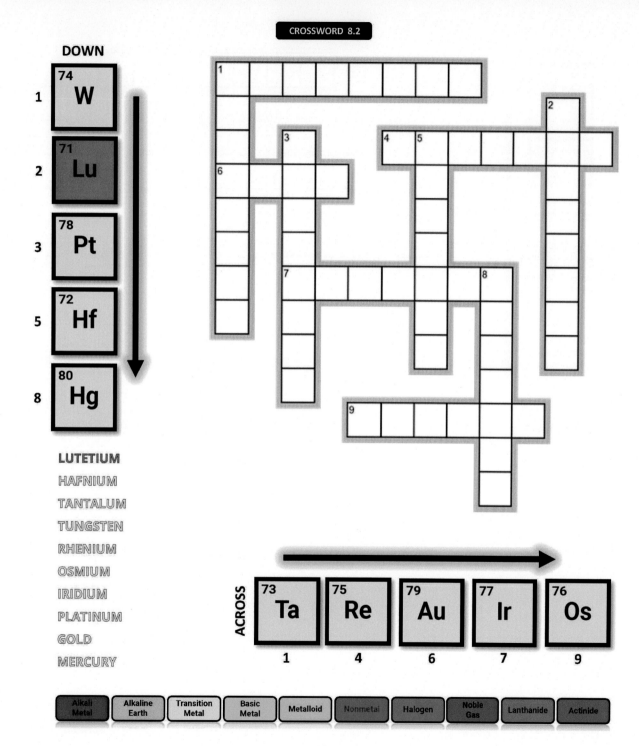

DOWN

1. ^{74} W
2. ^{71} Lu
3. ^{78} Pt
5. ^{72} Hf
8. ^{80} Hg

LUTETIUM
HAFNIUM
TANTALUM
TUNGSTEN
RHENIUM
OSMIUM
IRIDIUM
PLATINUM
GOLD
MERCURY

ACROSS

^{73} Ta — 1
^{75} Re — 4
^{79} Au — 6
^{77} Ir — 7
^{76} Os — 9

| Alkali Metal | Alkaline Earth | Transition Metal | Basic Metal | Metalloid | Nonmetal | Halogen | Noble Gas | Lanthanide | Actinide |

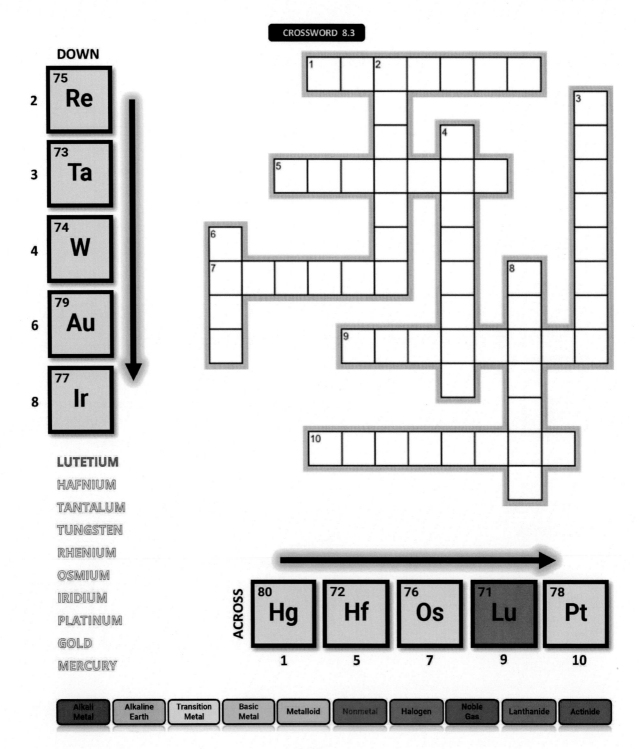

CROSSWORD 8.3

DOWN

2 | 75 Re
3 | 73 Ta
4 | 74 W
6 | 79 Au
8 | 77 Ir

LUTETIUM
HAFNIUM
TANTALUM
TUNGSTEN
RHENIUM
OSMIUM
IRIDIUM
PLATINUM
GOLD
MERCURY

ACROSS

80 Hg — 1
72 Hf — 5
76 Os — 7
71 Lu — 9
78 Pt — 10

Alkali Metal | Alkaline Earth | Transition Metal | Basic Metal | Metalloid | Nonmetal | Halogen | Noble Gas | Lanthanide | Actinide

38

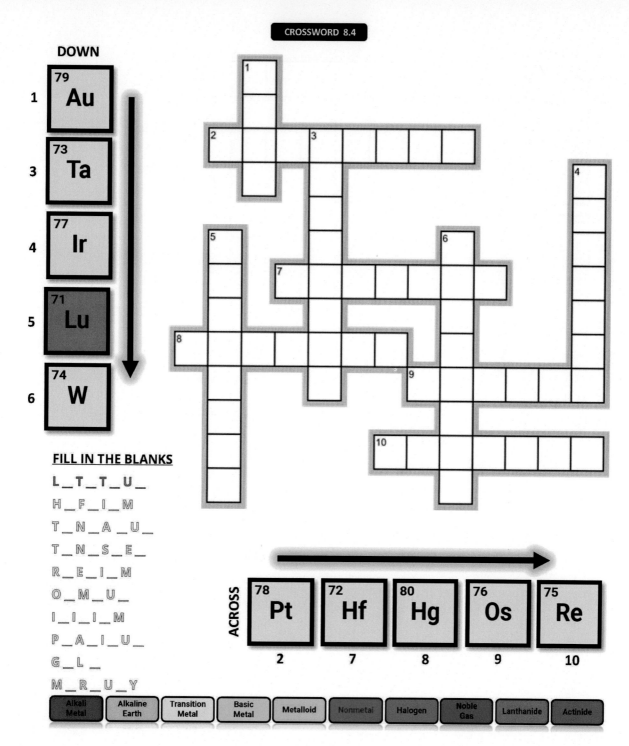

DOWN

1. ⁷⁹ Au
3. ⁷³ Ta
4. ⁷⁷ Ir
5. ⁷¹ Lu
6. ⁷⁴ W

FILL IN THE BLANKS

L _ T _ T _ U _
H _ F _ I _ M
T _ N _ A _ U _
T _ N _ S _ E _
R _ E _ I _ M
O _ M _ U _
I _ I _ I _ M
P _ A _ I _ U _
G _ L _
M _ R _ U _ Y

ACROSS

2. ⁷⁸ Pt
7. ⁷² Hf
8. ⁸⁰ Hg
9. ⁷⁶ Os
10. ⁷⁵ Re

| Alkali Metal | Alkaline Earth | Transition Metal | Basic Metal | Metalloid | Nonmetal | Halogen | Noble Gas | Lanthanide | Actinide |

CROSSWORD 8.5

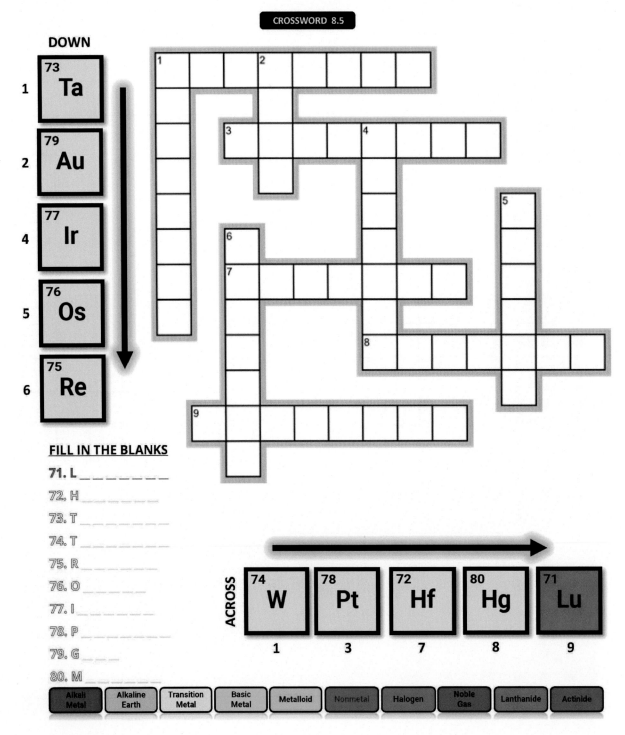

DOWN

1 | 73 Ta
2 | 79 Au
4 | 77 Ir
5 | 76 Os
6 | 75 Re

FILL IN THE BLANKS

71. L _ _ _ _ _ _ _ _ _

72. H _ _ _ _ _ _

73. T _ _ _ _ _ _ _

74. T _ _ _ _ _ _ _

75. R _ _ _ _ _ _

76. O _ _ _ _ _ _

77. I _ _ _ _ _ _

78. P _ _ _ _ _ _ _

79. G _ _ _

80. M _ _ _ _ _ _

ACROSS

74 W	78 Pt	72 Hf	80 Hg	71 Lu
1	3	7	8	9

Alkali Metal | Alkaline Earth | Transition Metal | Basic Metal | Metalloid | Nonmetal | Halogen | Noble Gas | Lanthanide | Actinide

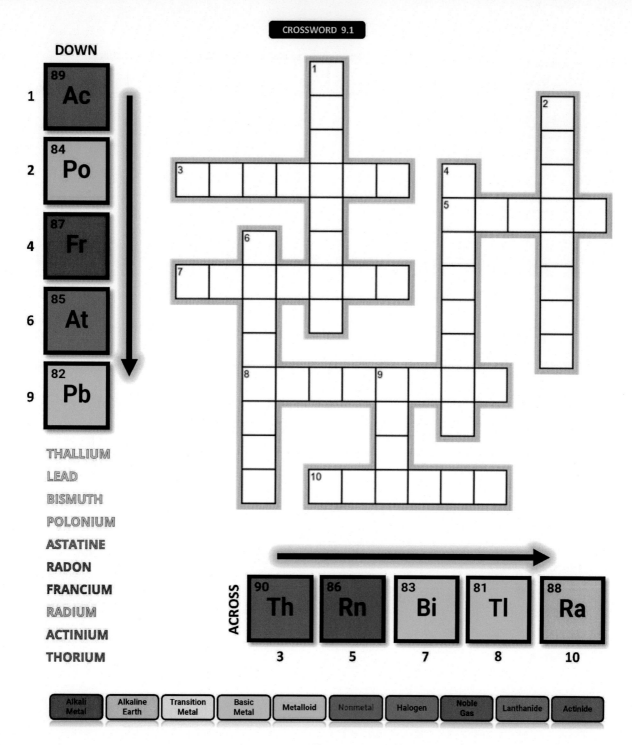

CROSSWORD 9.1

DOWN

1. 89 Ac
2. 84 Po
4. 87 Fr
6. 85 At
9. 82 Pb

THALLIUM
LEAD
BISMUTH
POLONIUM
ASTATINE
RADON
FRANCIUM
RADIUM
ACTINIUM
THORIUM

ACROSS

3. 90 Th
5. 86 Rn
7. 83 Bi
8. 81 Tl
10. 88 Ra

| Alkali Metal | Alkaline Earth | Transition Metal | Basic Metal | Metalloid | Nonmetal | Halogen | Noble Gas | Lanthanide | Actinide |

41

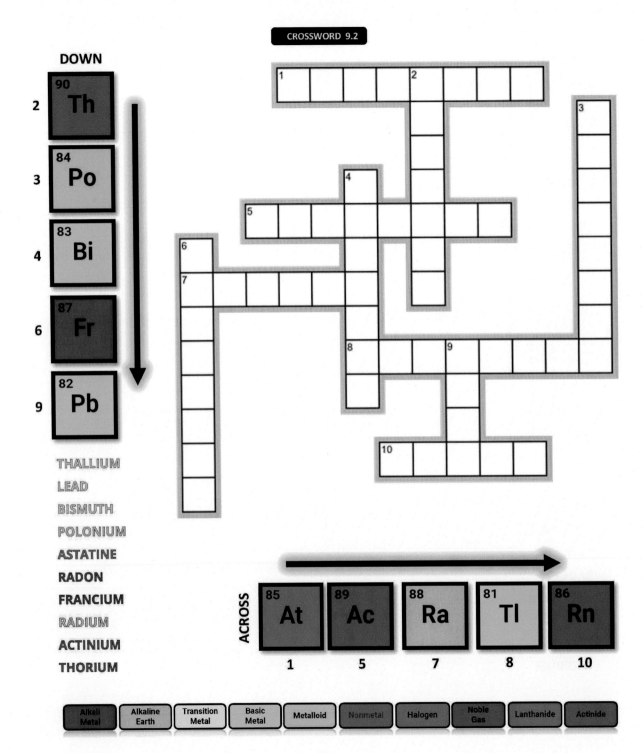

CROSSWORD 9.2

DOWN

2 — 90 Th
3 — 84 Po
4 — 83 Bi
6 — 87 Fr
9 — 82 Pb

THALLIUM
LEAD
BISMUTH
POLONIUM
ASTATINE
RADON
FRANCIUM
RADIUM
ACTINIUM
THORIUM

ACROSS

1 — 85 At
5 — 89 Ac
7 — 88 Ra
8 — 81 Tl
10 — 86 Rn

Alkali Metal | Alkaline Earth | Transition Metal | Basic Metal | Metalloid | Nonmetal | Halogen | Noble Gas | Lanthanide | Actinide

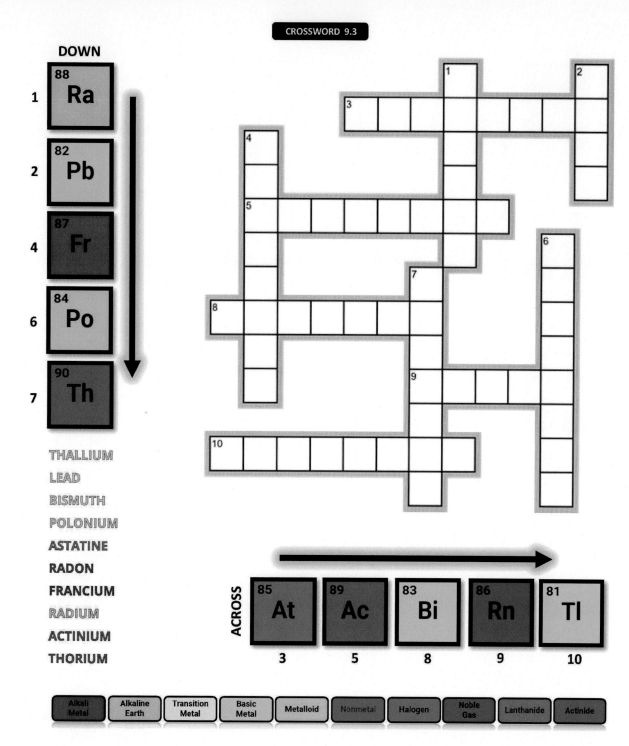

CROSSWORD 9.3

DOWN

1 — 88 Ra
2 — 82 Pb
4 — 87 Fr
6 — 84 Po
7 — 90 Th

THALLIUM
LEAD
BISMUTH
POLONIUM
ASTATINE
RADON
FRANCIUM
RADIUM
ACTINIUM
THORIUM

ACROSS

85 At — 3
89 Ac — 5
83 Bi — 8
86 Rn — 9
81 Tl — 10

Alkali Metal | Alkaline Earth | Transition Metal | Basic Metal | Metalloid | Nonmetal | Halogen | Noble Gas | Lanthanide | Actinide

CROSSWORD 9.4

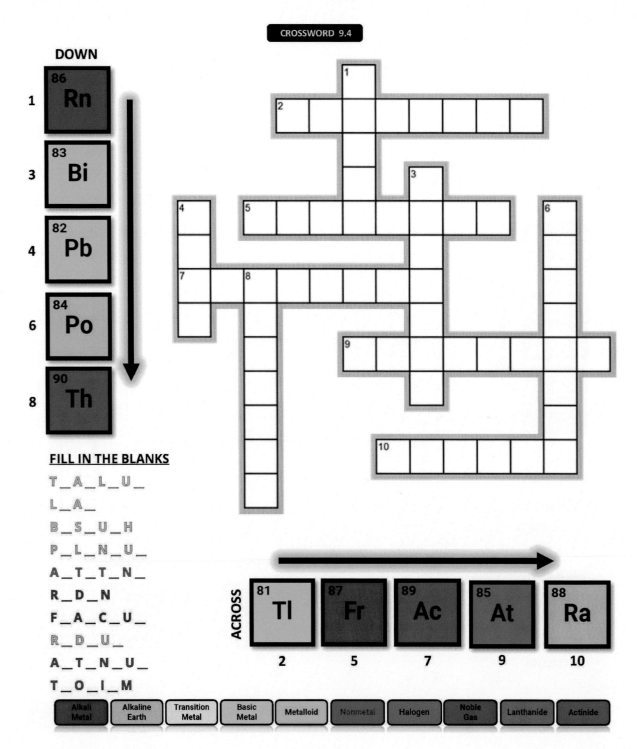

DOWN

1	**86 Rn**
3	**83 Bi**
4	**82 Pb**
6	**84 Po**
8	**90 Th**

FILL IN THE BLANKS

T _ A _ L _ U _

L _ A _

B _ S _ U _ H

P _ L _ N _ U _

A _ T _ T _ N _

R _ D _ N

F _ A _ C _ U _

R _ D _ U _

A _ T _ N _ U _

T _ O _ I _ M

ACROSS

81 Tl	87 Fr	89 Ac	85 At	88 Ra
2	5	7	9	10

Alkali Metal | Alkaline Earth | Transition Metal | Basic Metal | Metalloid | Nonmetal | Halogen | Noble Gas | Lanthanide | Actinide

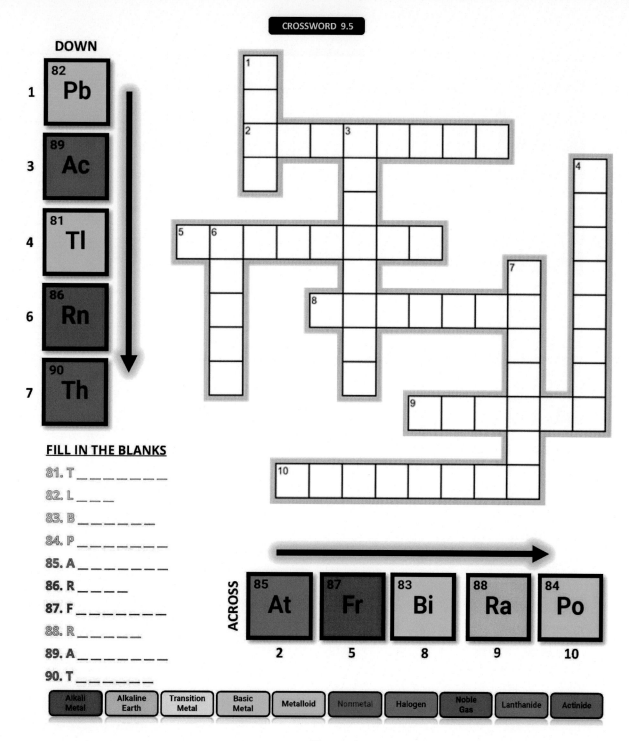

CROSSWORD 9.5

DOWN

1. 82 **Pb**
3. 89 **Ac**
4. 81 **Tl**
6. 86 **Rn**
7. 90 **Th**

FILL IN THE BLANKS

81. T _ _ _ _ _ _ _ _
82. L _ _ _ _
83. B _ _ _ _ _ _ _
84. P _ _ _ _ _ _ _
85. A _ _ _ _ _ _ _ _
86. R _ _ _ _ _
87. F _ _ _ _ _ _ _ _
88. R _ _ _ _ _ _
89. A _ _ _ _ _ _ _ _
90. T _ _ _ _ _ _ _

ACROSS

85 **At** — 2
87 **Fr** — 5
83 **Bi** — 8
88 **Ra** — 9
84 **Po** — 10

| Alkali Metal | Alkaline Earth | Transition Metal | Basic Metal | Metalloid | Nonmetal | Halogen | Noble Gas | Lanthanide | Actinide |

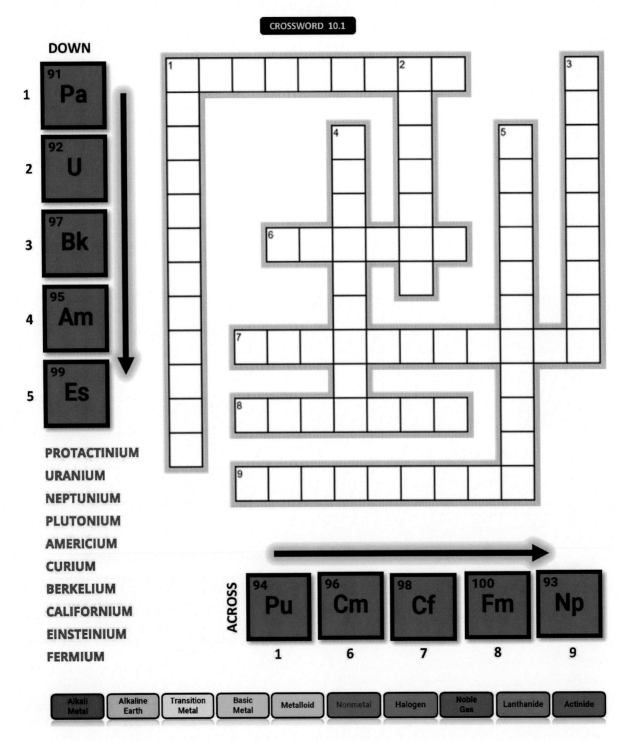

CROSSWORD 10.1

DOWN

1. **Pa** 91
2. **U** 92
3. **Bk** 97
4. **Am** 95
5. **Es** 99

PROTACTINIUM
URANIUM
NEPTUNIUM
PLUTONIUM
AMERICIUM
CURIUM
BERKELIUM
CALIFORNIUM
EINSTEINIUM
FERMIUM

ACROSS

1. **Pu** 94
6. **Cm** 96
7. **Cf** 98
8. **Fm** 100
9. **Np** 93

| Alkali Metal | Alkaline Earth | Transition Metal | Basic Metal | Metalloid | Nonmetal | Halogen | Noble Gas | Lanthanide | Actinide |

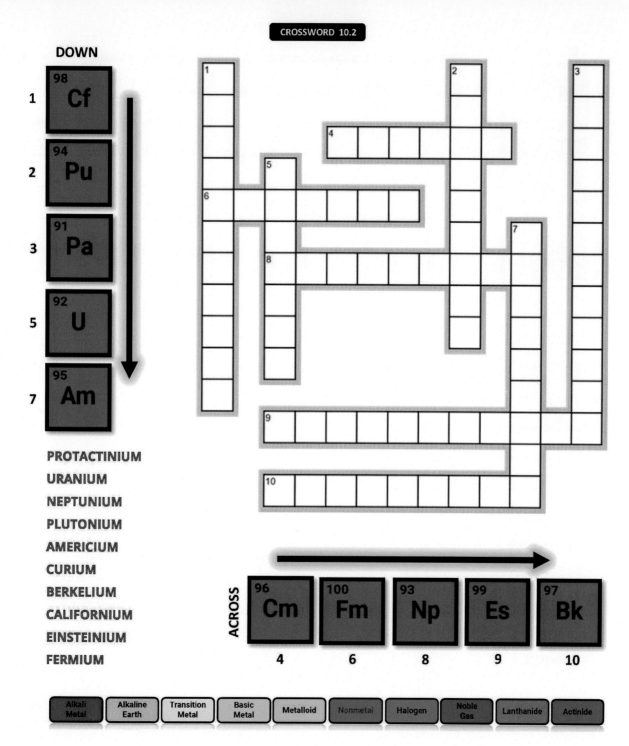

CROSSWORD 10.2

DOWN

1. ^{98} Cf
2. ^{94} Pu
3. ^{91} Pa
5. ^{92} U
7. ^{95} Am

PROTACTINIUM
URANIUM
NEPTUNIUM
PLUTONIUM
AMERICIUM
CURIUM
BERKELIUM
CALIFORNIUM
EINSTEINIUM
FERMIUM

ACROSS

^{96} Cm — 4
^{100} Fm — 6
^{93} Np — 8
^{99} Es — 9
^{97} Bk — 10

| Alkali Metal | Alkaline Earth | Transition Metal | Basic Metal | Metalloid | Nonmetal | Halogen | Noble Gas | Lanthanide | Actinide |

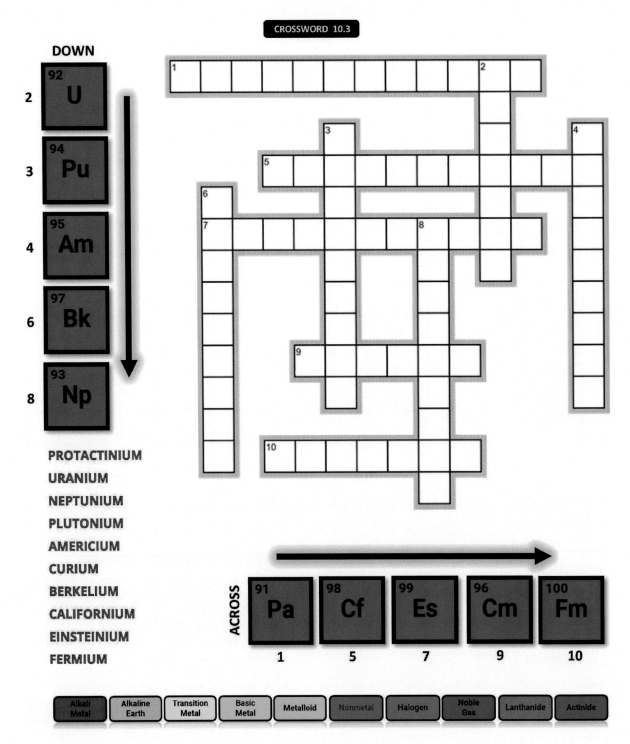

CROSSWORD 10.3

DOWN

2 — 92 U

3 — 94 Pu

4 — 95 Am

6 — 97 Bk

8 — 93 Np

PROTACTINIUM
URANIUM
NEPTUNIUM
PLUTONIUM
AMERICIUM
CURIUM
BERKELIUM
CALIFORNIUM
EINSTEINIUM
FERMIUM

ACROSS

91 Pa	98 Cf	99 Es	96 Cm	100 Fm
1	5	7	9	10

Alkali Metal	Alkaline Earth	Transition Metal	Basic Metal	Metalloid	Nonmetal	Halogen	Noble Gas	Lanthanide	Actinide

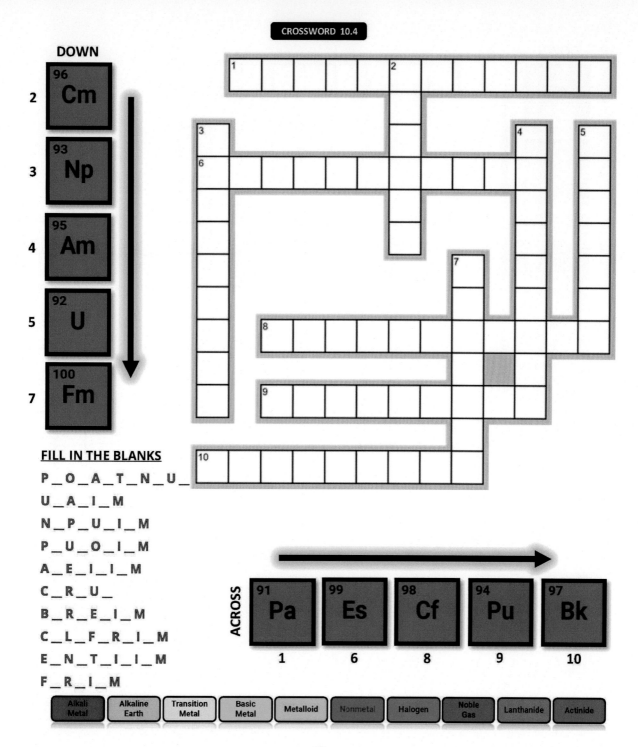

CROSSWORD 10.4

DOWN

2 | 96 Cm
3 | 93 Np
4 | 95 Am
5 | 92 U
7 | 100 Fm

FILL IN THE BLANKS

P _ O _ A _ T _ N _ U _

U _ A _ I _ M

N _ P _ U _ I _ M

P _ U _ O _ I _ M

A _ E _ I _ I _ M

C _ R _ U _

B _ R _ E _ I _ M

C _ L _ F _ R _ I _ M

E _ N _ T _ I _ I _ M

F _ R _ I _ M

ACROSS

91 Pa — 1
99 Es — 6
98 Cf — 8
94 Pu — 9
97 Bk — 10

Alkali Metal | Alkaline Earth | Transition Metal | Basic Metal | Metalloid | Nonmetal | Halogen | Noble Gas | Lanthanide | Actinide

49

DOWN

2. 91 Pa
3. 94 Pu
4. 97 Bk
5. 98 Cf
7. 95 Am

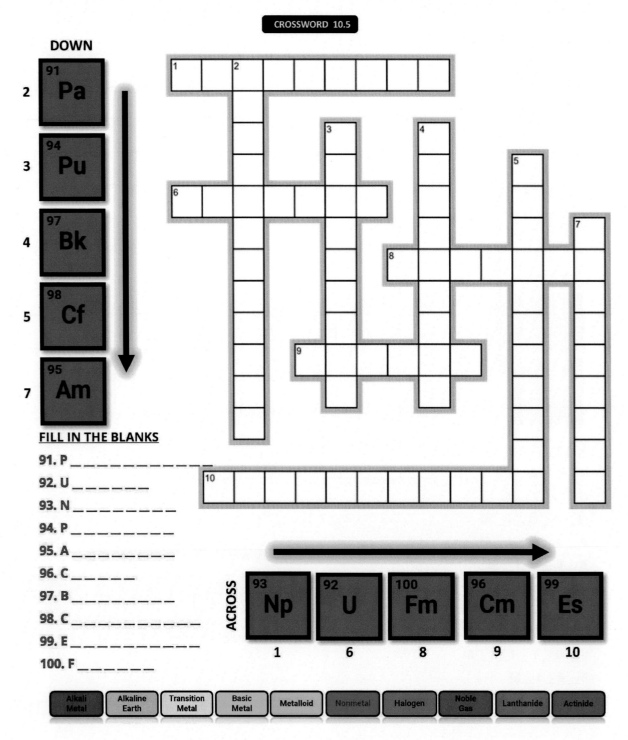

FILL IN THE BLANKS

91. P _ _ _ _ _ _ _ _ _ _ _ _ _
92. U _ _ _ _ _ _ _
93. N _ _ _ _ _ _ _ _ _
94. P _ _ _ _ _ _ _ _ _
95. A _ _ _ _ _ _ _ _ _
96. C _ _ _ _ _ _
97. B _ _ _ _ _ _ _ _ _
98. C _ _ _ _ _ _ _ _ _ _ _
99. E _ _ _ _ _ _ _ _ _ _ _
100. F _ _ _ _ _ _ _

ACROSS

93 Np — 1
92 U — 6
100 Fm — 8
96 Cm — 9
99 Es — 10

| Alkali Metal | Alkaline Earth | Transition Metal | Basic Metal | Metalloid | Nonmetal | Halogen | Noble Gas | Lanthanide | Actinide |

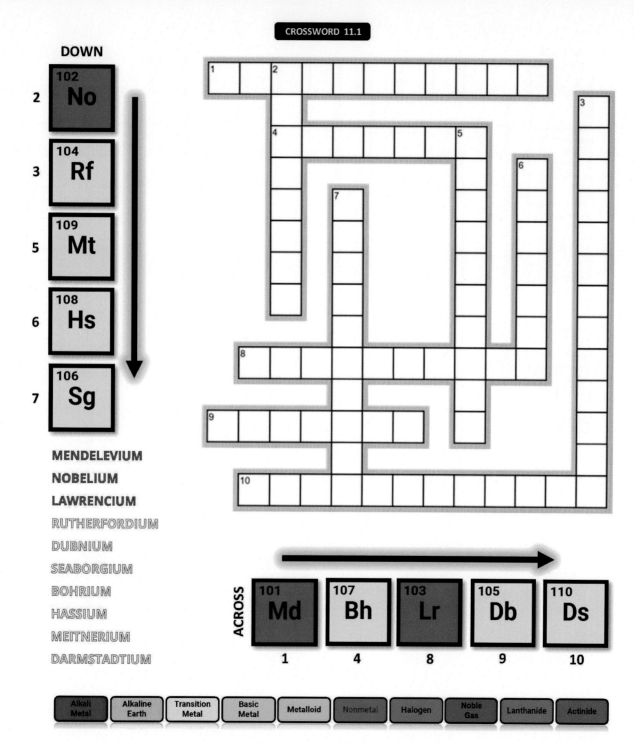

CROSSWORD 11.1

DOWN

2 — 102 No
3 — 104 Rf
5 — 109 Mt
6 — 108 Hs
7 — 106 Sg

MENDELEVIUM
NOBELIUM
LAWRENCIUM
RUTHERFORDIUM
DUBNIUM
SEABORGIUM
BOHRIUM
HASSIUM
MEITNERIUM
DARMSTADTIUM

ACROSS

1 — 101 Md
4 — 107 Bh
8 — 103 Lr
9 — 105 Db
10 — 110 Ds

Alkali Metal | Alkaline Earth | Transition Metal | Basic Metal | Metalloid | Nonmetal | Halogen | Noble Gas | Lanthanide | Actinide

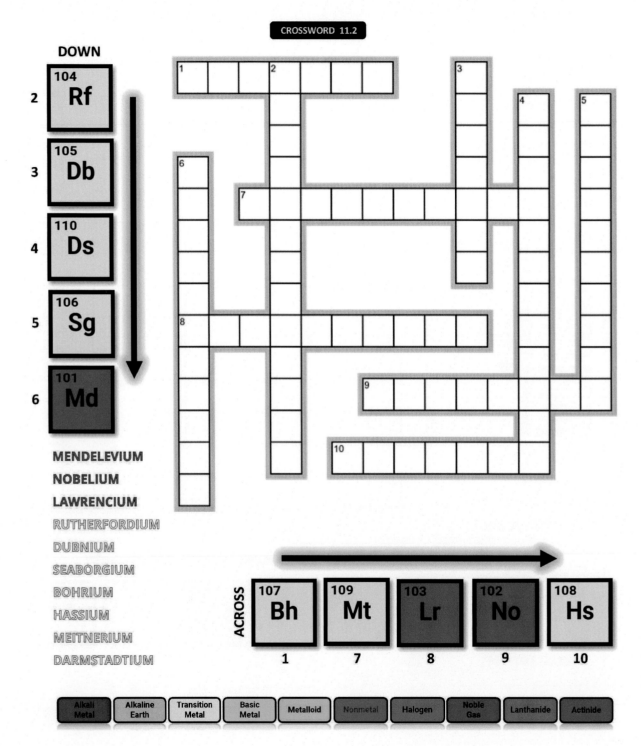

CROSSWORD 11.2

DOWN

2. 104 **Rf**
3. 105 **Db**
4. 110 **Ds**
5. 106 **Sg**
6. 101 **Md**

MENDELEVIUM
NOBELIUM
LAWRENCIUM
RUTHERFORDIUM
DUBNIUM
SEABORGIUM
BOHRIUM
HASSIUM
MEITNERIUM
DARMSTADTIUM

ACROSS

1. 107 **Bh**
7. 109 **Mt**
8. 103 **Lr**
9. 102 **No**
10. 108 **Hs**

| Alkali Metal | Alkaline Earth | Transition Metal | Basic Metal | Metalloid | Nonmetal | Halogen | Noble Gas | Lanthanide | Actinide |

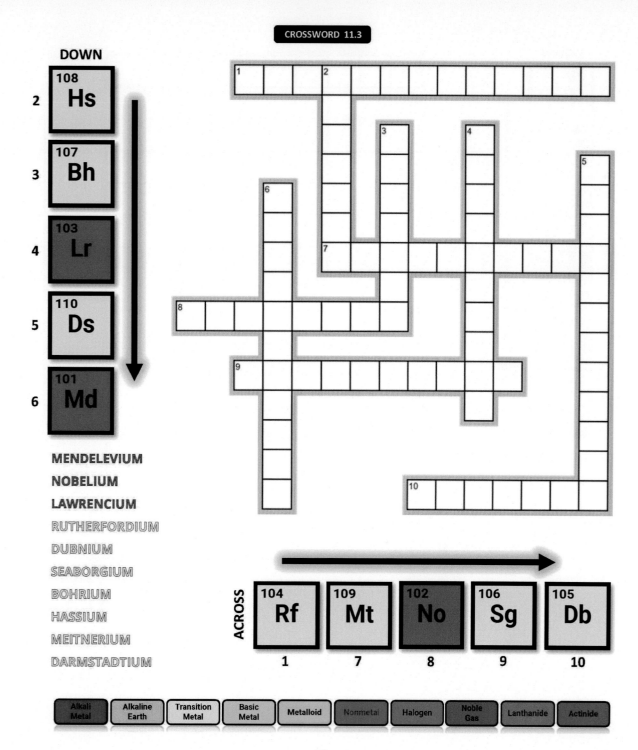

CROSSWORD 11.3

DOWN

2. 108 Hs
3. 107 Bh
4. 103 Lr
5. 110 Ds
6. 101 Md

MENDELEVIUM
NOBELIUM
LAWRENCIUM
RUTHERFORDIUM
DUBNIUM
SEABORGIUM
BOHRIUM
HASSIUM
MEITNERIUM
DARMSTADTIUM

ACROSS

1. 104 Rf
7. 109 Mt
8. 102 No
9. 106 Sg
10. 105 Db

Alkali Metal | Alkaline Earth | Transition Metal | Basic Metal | Metalloid | Nonmetal | Halogen | Noble Gas | Lanthanide | Actinide

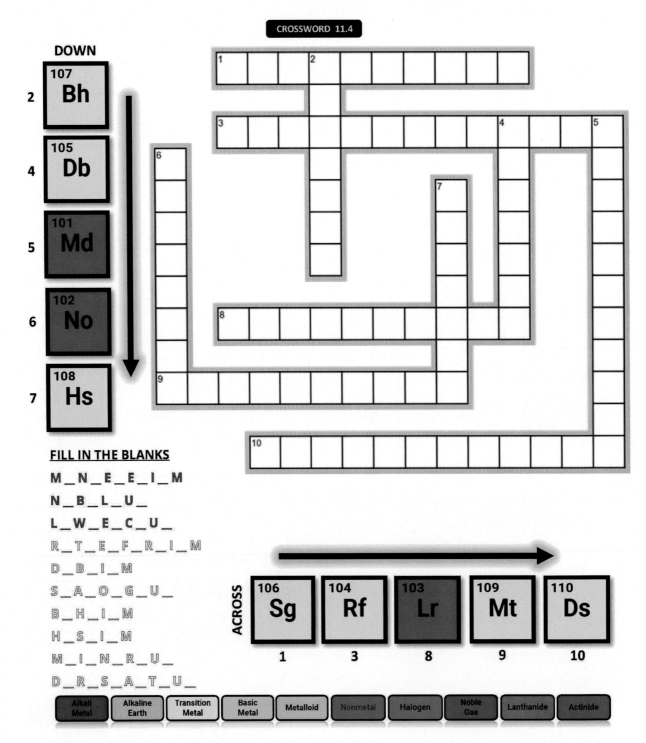

CROSSWORD 11.4

DOWN

#		
2	107 Bh	
4	105 Db	
5	101 Md	
6	102 No	
7	108 Hs	

FILL IN THE BLANKS

M _ N _ E _ I _ M

N _ B _ L _ U _

L _ W _ E _ C _ U _

R _ T _ E _ F _ R _ I _ M

D _ B _ I _ M

S _ A _ O _ G _ U _

B _ H _ I _ M

H _ S _ I _ M

M _ I _ N _ R _ U _

D _ R _ S _ A _ T _ U _

ACROSS

106 Sg	104 Rf	103 Lr	109 Mt	110 Ds
1	3	8	9	10

Alkali Metal | Alkaline Earth | Transition Metal | Basic Metal | Metalloid | Nonmetal | Halogen | Noble Gas | Lanthanide | Actinide

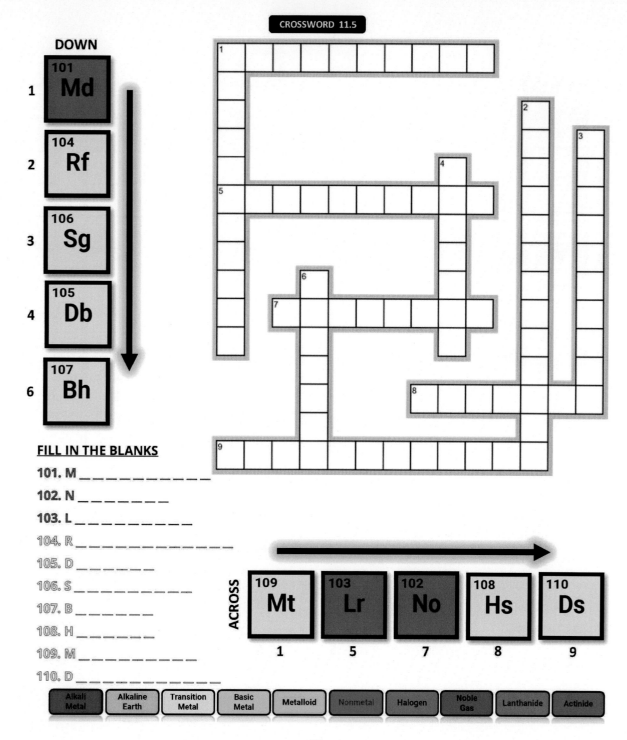

CROSSWORD 11.5

DOWN

1 101 Md
2 104 Rf
3 106 Sg
4 105 Db
6 107 Bh

FILL IN THE BLANKS

101. M _ _ _ _ _ _ _ _ _ _ _ _

102. N _ _ _ _ _ _ _ _

103. L _ _ _ _ _ _ _ _ _ _

104. R _ _ _ _ _ _ _ _ _ _ _ _ _

105. D _ _ _ _ _ _

106. S _ _ _ _ _ _ _ _ _

107. B _ _ _ _ _ _

108. H _ _ _ _ _ _

109. M _ _ _ _ _ _ _ _ _

110. D _ _ _ _ _ _ _ _ _ _

ACROSS

109 Mt — 1
103 Lr — 5
102 No — 7
108 Hs — 8
110 Ds — 9

Alkali Metal | Alkaline Earth | Transition Metal | Basic Metal | Metalloid | Nonmetal | Halogen | Noble Gas | Lanthanide | Actinide

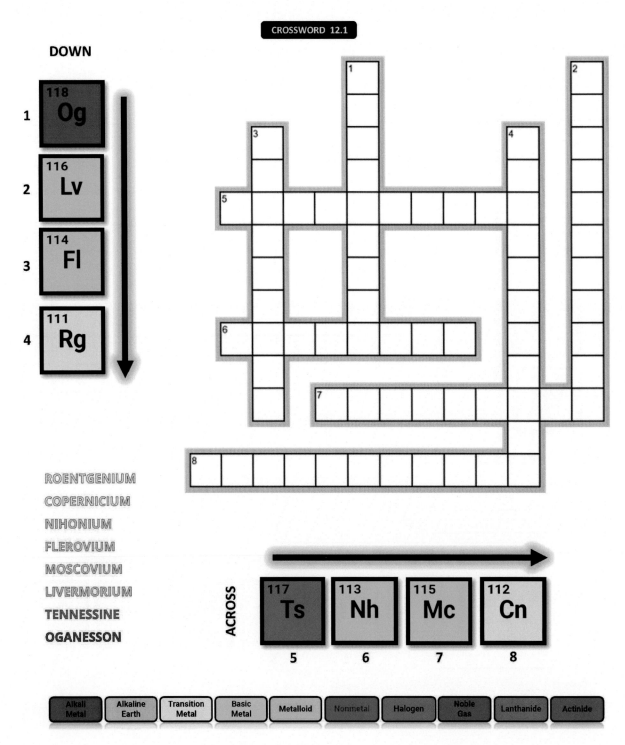

CROSSWORD 12.1

DOWN

1 | 118 Og
2 | 116 Lv
3 | 114 Fl
4 | 111 Rg

ROENTGENIUM
COPERNICIUM
NIHONIUM
FLEROVIUM
MOSCOVIUM
LIVERMORIUM
TENNESSINE
OGANESSON

ACROSS

| 117 Ts | 113 Nh | 115 Mc | 112 Cn |
| 5 | 6 | 7 | 8 |

| Alkali Metal | Alkaline Earth | Transition Metal | Basic Metal | Metalloid | Nonmetal | Halogen | Noble Gas | Lanthanide | Actinide |

56

DOWN

1 | 112 **Cn**

2 | 117 **Ts**

3 | 113 **Nh**

5 | 115 **Mc**

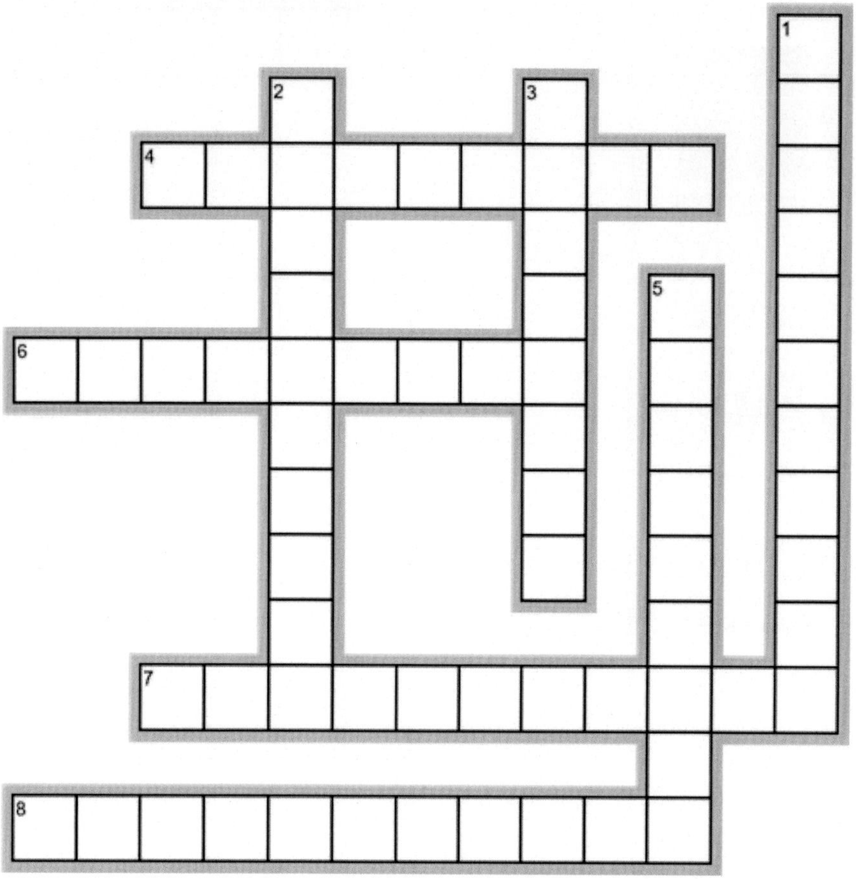

ROENTGENIUM
COPERNICIUM
NIHONIUM
FLEROVIUM
MOSCOVIUM
LIVERMORIUM
TENNESSINE
OGANESSON

ACROSS

114 **Fl** — 4
118 **Og** — 6
111 **Rg** — 7
116 **Lv** — 8

| Alkali Metal | Alkaline Earth | Transition Metal | Basic Metal | Metalloid | Nonmetal | Halogen | Noble Gas | Lanthanide | Actinide |

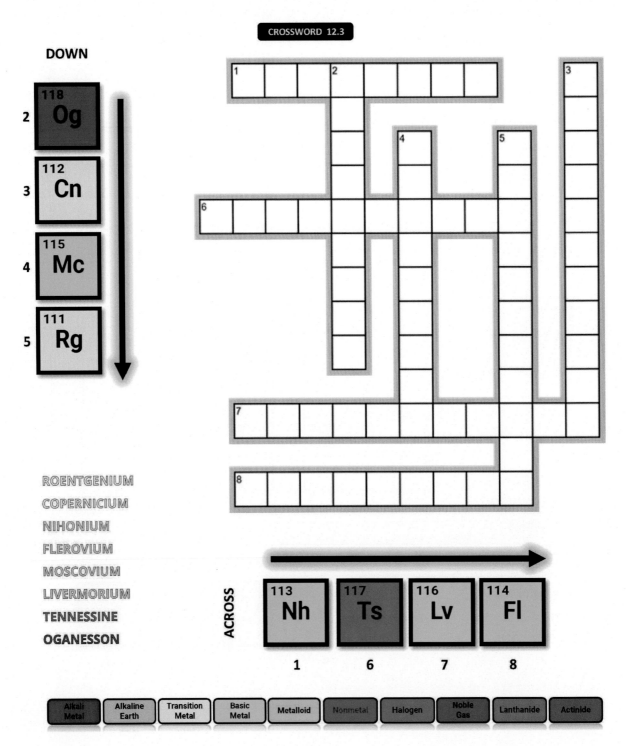

DOWN

2 | 118 Og
3 | 112 Cn
4 | 115 Mc
5 | 111 Rg

ROENTGENIUM
COPERNICIUM
NIHONIUM
FLEROVIUM
MOSCOVIUM
LIVERMORIUM
TENNESSINE
OGANESSON

ACROSS

113 Nh — 1
117 Ts — 6
116 Lv — 7
114 Fl — 8

Alkali Metal | Alkaline Earth | Transition Metal | Basic Metal | Metalloid | Nonmetal | Halogen | Noble Gas | Lanthanide | Actinide

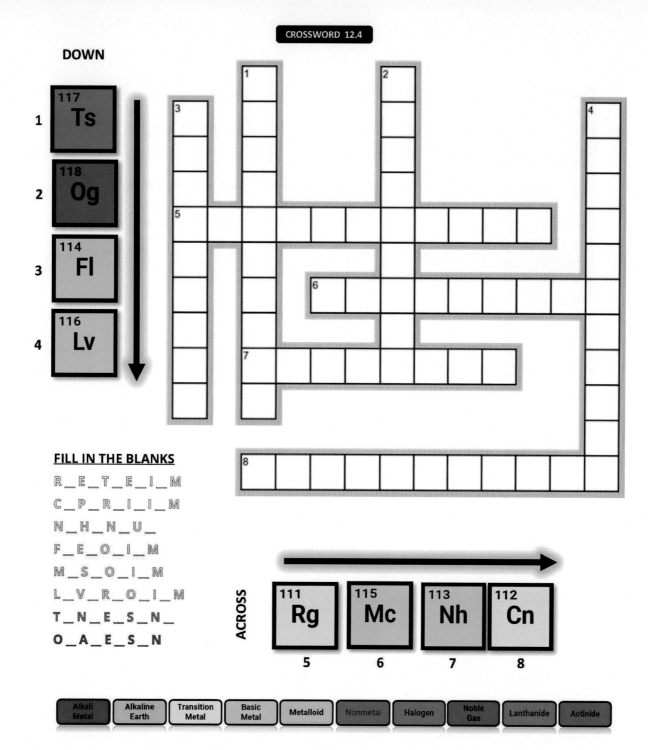

CROSSWORD 12.4

DOWN

1 — 117 Ts
2 — 118 Og
3 — 114 Fl
4 — 116 Lv

FILL IN THE BLANKS

R _ E _ T _ E _ I _ M
C _ P _ R _ I _ I _ M
N _ H _ N _ U _
F _ E _ O _ I _ M
M _ S _ O _ I _ M
L _ V _ R _ O _ I _ M
T _ N _ E _ S _ N _
O _ A _ E _ S _ N

ACROSS

5 — 111 Rg
6 — 115 Mc
7 — 113 Nh
8 — 112 Cn

Alkali Metal | Alkaline Earth | Transition Metal | Basic Metal | Metalloid | Nonmetal | Halogen | Noble Gas | Lanthanide | Actinide

DOWN

1 112 **Cn**

2 113 **Nh**

3 117 **Ts**

4 111 **Rg**

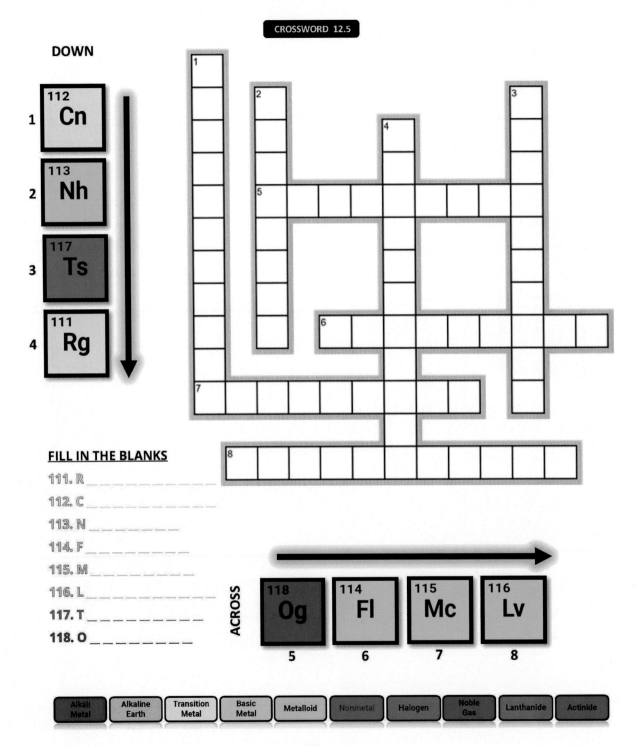

FILL IN THE BLANKS

111. R _ _ _ _ _ _ _ _ _ _

112. C _ _ _ _ _ _ _ _ _ _ _ _

113. N _ _ _ _ _ _ _ _ _

114. F _ _ _ _ _ _ _ _ _

115. M _ _ _ _ _ _ _ _ _

116. L _ _ _ _ _ _ _ _ _ _ _

117. T _ _ _ _ _ _ _ _ _

118. O _ _ _ _ _ _ _ _ _

ACROSS

118 **Og** 5

114 **Fl** 6

115 **Mc** 7

116 **Lv** 8

Alkali Metal | Alkaline Earth | Transition Metal | Basic Metal | Metalloid | Nonmetal | Halogen | Noble Gas | Lanthanide | Actinide

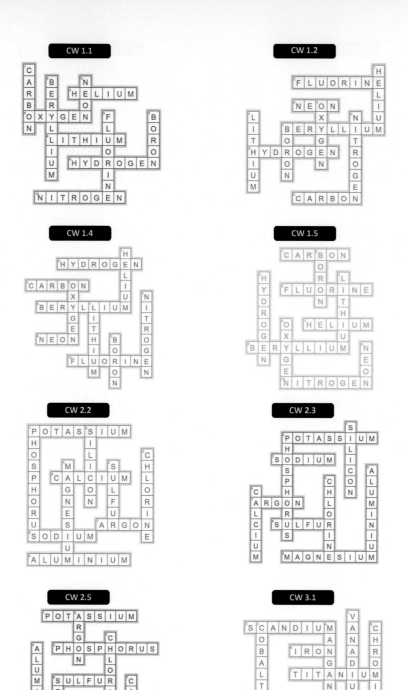
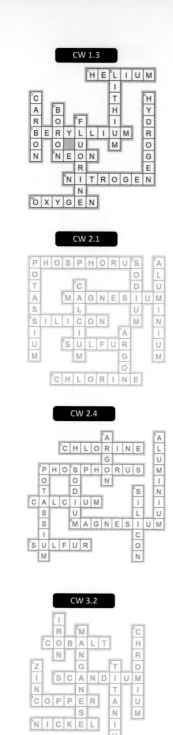

61

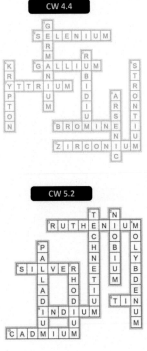

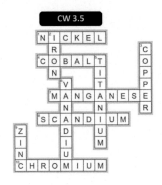
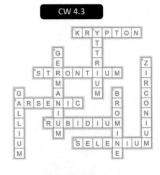

CW 3.3

CW 3.4

CW 3.5

CW 4.1

CW 4.2

CW 4.3

CW 4.4

CW 4.5

CW 5.1

CW 5.2

CW 5.3

CW 5.4

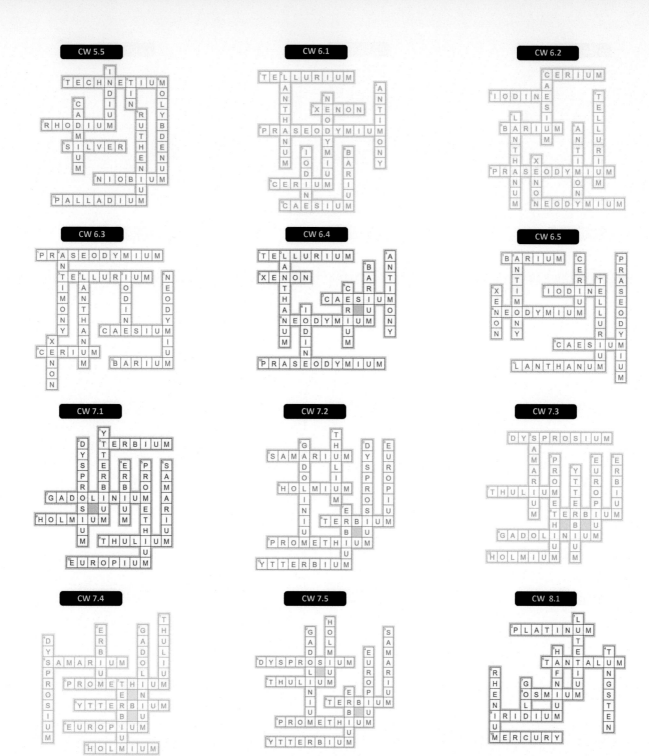

63

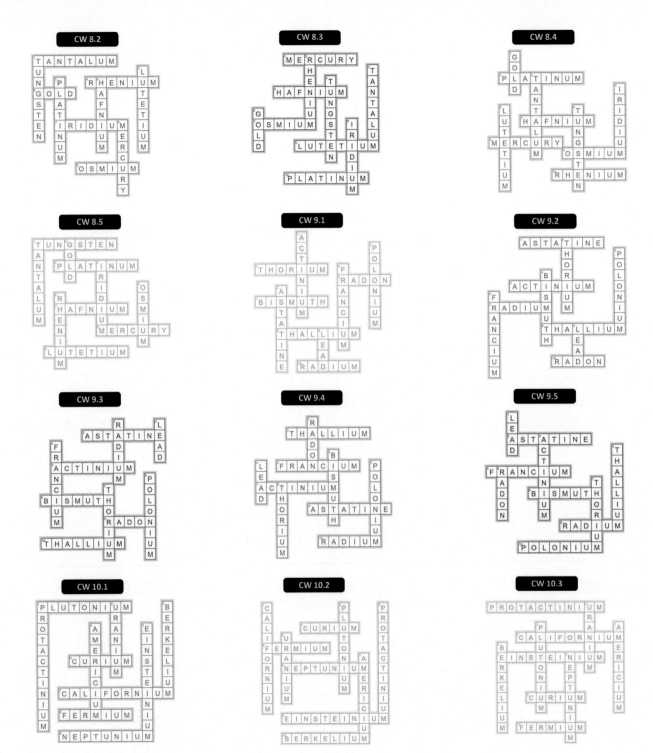

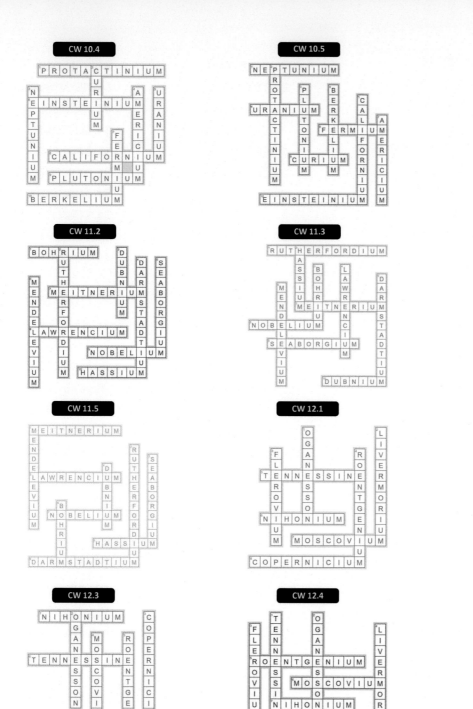

1. HYDROGEN	11. SODIUM	21. SCANDIUM	31. GALLIUM
2. HELIUM	12. MAGNESIUM	22. TITANIUM	32. GERMANIUM
3. LITHIUM	13. ALUMINIUM	23. VANADIUM	33. ARSENIC
4. BERYLLIUM	14. SILICON	24. CHROMIUM	34. SELENIUM
5. BORON	15. PHOSPHORUS	25. MANGANESE	35. BROMINE
6. CARBON	16. SULFUR	26. IRON	36. KRYPTON
7. NITROGEN	17. CHLORINE	27. COBALT	37. RUBIDIUM
8. OXYGEN	18. ARGON	28. NICKEL	38. STRONTIUM
9. FLUORINE	19. POTASSIUM	29. COPPER	39. YTTRIUM
10. NEON	20. CALCIUM	30. ZINC	40. ZIRCONIUM

41. NIOBIUM	51. ANTIMONY	61. PROMETHIUM	71. LUTETIUM
42. MOLYBDENUM	52. TELLURIUM	62. SAMARIUM	72. HAFNIUM
43. TECHNETIUM	53. IODINE	63. EUROPIUM	73. TANTALUM
44. RUTHENIUM	54. XENON	64. GADOLINIUM	74. TUNGSTEN
45. RHODIUM	55. CAESIUM	65. TERBIUM	75. RHENIUM
46. PALLADIUM	56. BARIUM	66. DYSPROSIUM	76. OSMIUM
47. SILVER	57. LANTHANUM	67. HOLMIUM	77. IRIDIUM
48. CADMIUM	58. CERIUM	68. ERBIUM	78. PLATINUM
49. INDIUM	59. PRASEODYMIUM	69. THULIUM	79. GOLD
50. TIN	60. NEODYMIUM	70. YTTERBIUM	80. MERCURY

81. THALLIUM	91. PROTACTINIUM	101. MENDELEVIUM	111. ROENTGENIUM
82. LEAD	92. URANIUM	102. NOBELIUM	112. COPERNICIUM
83. BISMUTH	93. NEPTUNIUM	103. LAWRENCIUM	113. NIHONIUM
84. POLONIUM	94. PLUTONIUM	104. RUTHERFORDIUM	114. FLEROVIUM
85. ASTATINE	95. AMERICIUM	105. DUBNIUM	115. MOSCOVIUM
86. RADON	96. CURIUM	106. SEABORGIUM	116. LIVERMORIUM
87. FRANCIUM	97. BERKELIUM	107. BOHRIUM	117. TENNESSINE
88. RADIUM	98. CALIFORNIUM	108. HASSIUM	118. OGANESSON
89. ACTINIUM	99. EINSTEINIUM	109. MEITNERIUM	
90. THORIUM	100. FERMIUM	110. DARMSTADTIUM	